Lecture Notes in Computer Science 10127

Commenced Publication in 1973
Founding and Former Series Editors:
Gerhard Goos, Juris Hartmanis, and Jan van Leeuwen

More information about this series at http://www.springer.com/series/7409

Anupam Basu · Sukhendu Das
Patrick Horain · Samit Bhattacharya (Eds.)

Intelligent Human Computer Interaction

8th International Conference, IHCI 2016
Pilani, India, December 12–13, 2016
Proceedings

Springer

Editors
Anupam Basu
IIT Kharagpur
Kharagpur
India

Sukhendu Das
IIT Madras
Chennai
India

Patrick Horain
Telecom SudParis
Evry
France

Samit Bhattacharya
IIT Guwahati
Guwahati
India

ISSN 0302-9743   ISSN 1611-3349 (electronic)
Lecture Notes in Computer Science
ISBN 978-3-319-52502-0   ISBN 978-3-319-52503-7   (eBook)
DOI 10.1007/978-3-319-52503-7

Library of Congress Control Number: 2016963649

LNCS Sublibrary: SL3 – Information Systems and Applications, incl. Internet/Web, and HCI

Printed on acid-free paper

This Springer imprint is published by Springer Nature
The registered company is Springer International Publishing AG
The registered company address is: Gewerbestrasse 11, 6330 Cham, Switzerland

Preface

This volume contains the papers presented at the 8th International Conference on Intelligent Human Computer Interaction, held during December 12–13, 2016, in Pilani, Rajasthan, India. Three leading research and academic institutes of India, namely, the Council of Scientific and Industrial Research — Central Electronics Engineering Research Institute Pilani (CSIR-CEERI), the Birla Institute of Technology and Science Pilani (BITS-Pilani), and the Indian Institute of Information Technology, Allahabad (IIIT-Allahabad), jointly organized the conference in Pilani, Rajasthan, India. The conference, which had its first edition on 2008, has as its focus the research issues related to human–computer interaction. This year, the eighth edition of the conference was successful in attracting the academics, researchers, and industry practitioners from all over the globe.

The current volume contains the selected papers from 115 submissions from all over the world. The Program Committee (PC) accepted 22 full papers for oral and seven papers for poster presentations after a rigorous review process with multiple reviews for each paper. Finally, 22 camera-ready papers were received for oral presentation. We thank all the expert reviewers for their invaluable support and constructive review comments. We are grateful to the PC members who made enormous efforts in reviewing and selecting the papers. Without the untiring efforts of the PC members, the reviewers, and the contributions of the authors, the conference would not have been possible.

The entire process of submission, refereeing, e-meetings of the PC for selecting the papers, and compiling the proceedings was done through Microsoft's conference management tool and special thanks go to the Microsoft team for providing such a highly configurable conference management system.

One of the hallmarks of the IHCI conference series is the high quality of the plenary/invited presentations. This year, we were fortunate to have the invited talks of three eminent speakers, namely, Prof. Thomas Ertl (VISUS, University of Stuttgart, Germany), Dr. Manas K. Mandal (Director General - Life Sciences, DRDO, India), and Prof. Sudeep Sarkar (University of South Florida, USA). It is indeed a great pleasure for us to thank the invited speakers who agreed to present at the conference coming from far off places in mid-December. We are grateful to them for their time and efforts.

We thank all the members of the Organizing Committee for making all the arrangements for the conference. We are grateful to CSIR-CEERI, Pilani, and BITS-Pilani for all the assistance provided for running the conference. In particular, Prof. Santanu Chaudhury (Director, CSIR-CEERI, Pilani) and Prof. A.K. Sarkar (Director, BITS-Pilani, Pilani Campus) helped us tremendously at key points with the logistics of running the PC meeting. We are thankful to Mr. Dhiraj (CSIR-CEERI, Pilani), Mr. Pramod Tanwar (CSIR-CEERI, Pilani), Mr. PVL Reddy (CSIR-CEERI, Pilani), and Dr. Pankaj Vyas (BITS-Pilani) for their enormous efforts in compiling and formatting the final manuscripts of authors as per the LNCS guidelines. We are also grateful to all the graduate as well as under graduate students of BITS-Pilani and the young scientists, PGRP students and project staffs of CSIR-CEERI Pilani and other staff members of both the institutes for their assistance in making necessary arrangements for the smooth conduct of the conference.

Finally, yet importantly, thanks go to Alfred Hofmann (Associate Director Publishing, Springer) for readily agreeing to publish the proceedings in the LNCS series. Thanks also go to his team and in particular to Ms. Suvira Srivastava (Springer India) in preparing the proceedings meticulously for publication.

December 2016   Anupam Basu

Organizers

General Chair

Anupam Basu   IIT Kharagpur, India

Program Chairs

Sukhendu Das   IIT Madras, India
Patrick Horain   Telecom SudParis, France
Samit Bhattacharya   IIT Guwahati, India

Organizing Chairs

Atanendu S. Mandal   CSIR-CEERI Pilani, India
Abhijit R. Asati   BITS Pilani, India
Yashvardhan Sharma   BITS Pilani, India

Publication Chairs

Dhiraj   CSIR-CEERI Pilani, India
Pankaj Vyas   BITS Pilani, India
Pramod Kumar Tanwar   CSIR-CEERI Pilani, India

Industrial Liaison Chairs

S.A. Akbar   CSIR-CEERI Pilani, India
Avinash Gautam   BITS Pilani, India

Publicity Chairs

P.V.L. Reddy   CSIR-CEERI Pilani, India
Abhijit R. Asati   BITS Pilani, India

Advisory Committee

Souvik Bhattacharyya   VC, BITS Pilani, India
Santanu Chaudhury   CSIR-CEERI Pilani, India
Ashoke K. Sarkar   BITS Pilani, India
Somnath Biswas   IIIT Allahabad, India
Uma Shankar Tiwary   IIIT Allahabad, India
Chandra Shekhar   BITS Pilani, and CSIR-CEERI Pilani, India

Jaroslav Pokorný	Charles University, Czech Republic
Ling Guan	Ryerson University, Canada

Financial Chair

Subash Chandra Bose	CSIR-CEERI Pilani, India

Plenary Chair

J.L. Raheja	CSIR-CEERI Pilani, India
Lavika Goel	BITS Pilani, India

Plenary Speakers

Thomas Ertl	VISUS, University of Stuttgart, Germany
Manas K. Mandal	Life Sciences (DGLS), DRDO, India
Sudeep Sarkar	University of South Florida, USA

Industrial Session Chair (Special Session)

Lipika Dey	Innovation Labs, Tata Consultancy Services, New Delhi, India

Program Committee

Abhijit Karmakar	CSIR-CEERI Pilani, India
Alok Kanti Deb	IIT Kharagpur, India
Amine Chellali	Evry Val d'Essonne University, France
Amit Konar	Jadavpur University, India
Amrita Basu	Jadavpur University, India
Anirudha Joshi	IIT Bombay India
Anupam Basu	IIT Kharagpur, India
Atanendu S Mandal	CSIR-CEERI Pilani, India
Catherine Achard	University Pierre et Marie Curie, France
Dakshina Ranjan Kisku	NIT Durgapur, India
Daniel Wesierski	Gdańsk University of Technology, Poland
David Antonio Gòmez Jàuregui	LIMSI, France
Debasis Samanta	IIT Kharagpur, India
Debasish Mazumdar	CDAC Kolkata, India
Ehtesham Hassan	Tata Consultancy Services, India
Ekram Khan	Aligarh Muslim University, India
Gaurav Harit	IIT Jodhpur, India
Geehyuk Lee	KAIST, South Korea
Harish Karnick	IIT Kanpur, India
Jagriti P. Galphade	National Institute of Design, India

Jan Platoš	TU Ostrava, Czech Republic
Jayadeva	IIT Delhi, India
Keith Cheverst	Lancaster University, UK
Kuntal Ghosh	ISI Kolkata, India
Laxmidhar Behera	IIT Kanpur, India
Lipika Dey	Tata Consultancy Services, India
K. Madhava Krishna	IIIT Hyderabad, India
Malik Mallem	Université d'Évry-Val-d'Essonne, Évry
M. Manivannan	IIT Madras, India
Manuel Montes-y-Gòmez	National Institute of Astrophysics, Mexico
Mirza Mohd. Sufyan Beg	Jamia Millia University, India
Narayanan Srinivasan	CBCS, Allahabad University, India
Partha Pratim Das	IIT Kharagpur, India
Patrick Horain	Telecom SudParis, France
Plaban K. Bhowmick	IIT Kharagpur, India
Prem C. Pandey	IIT Bombay, India
Rahul Banerjee	BITS Pilani, India
Ratna Sanyal	IIIT Allahabad, India
B. Ravindran	IIT Madras, India
Samit Bhattacharya	IIT Guwahati, India
Sanasam Singh	IIT Guwahati, India
Santanu Chaudhury	CSIR-CEERI Pilani, India
Subhasis Chaudhuri	IIT Bombay, India
Sudip Sanyal	IIIT, Allahabad, India
Sukhendu Das	IIT Madras, India
Sumeet Agarwal	IIT Delhi, India
Santanu Chakraborti	IIT Madras, India
Tamarapalli Venkatesh	IIT Guwahati, India
Tanveer Siddiqui	University of Allahabad, India
Tapan Gandhi	IIT Delhi, India
U.S. Tiwary	IIIT Allahabad, India

Organizing Committee

CSIR-CEERI, Pilani

Santanu Chaudhury (Chair)	CSIR-CEERI Pilani, India
Raj Singh	CSIR-CEERI Pilani, India
P. Bhanu Prasad	CSIR-CEERI Pilani, India
S.A. Akbar	CSIR-CEERI Pilani, India
K.P. Sharma	CSIR-CEERI Pilani, India
P.V.L. Reddy	CSIR-CEERI Pilani, India
Atanendu S. Mandal (Convener)	CSIR-CEERI Pilani, India

BITS Pilani, Pilani Campus

Ashoke K. Sarkar (Co-chair)	BITS Pilani, India
S.C. Sivasubramanian	BITS Pilani, India
Rahul Banerjee	BITS Pilani, India
Anu Gupta	BITS Pilani, India
Surekha Bhanot	BITS Pilani, India
J.P. Mishra	BITS Pilani, India
Yashvardhan Sharma (Co-convener)	BITS Pilani, India

Invited Talks

Interactive Visualization: A Key Discipline for Big Data Analysis

Thomas Etrl

Universittsstrae 38, 70569 Stuttgart, Germany
Thomas.Ertl@vis.uni-stuttgart.de
http://www.vis.uni-stuttgart.
de/institut/mitarbeiter/thomas-ertl/

Abstract. Big Data has become the general term relating to the benefits and threats which result from the huge amount of data collected in all parts of society. While data acquisition, storage and access are relevant technical aspects, the analysis of the collected data turns out to be at the core of the Big Data challenge. Automatic data mining and information retrieval techniques have made much progress but many application scenarios remain in which the human in the loop plays an essential role. Consequently interactive visualization techniques have become a key discipline of Big Data analysis and the field is reaching out to many new application domains. This talk will give examples from current visualization research projects at the University of Stuttgart demonstrating the thematic breadth of application scenarios and the technical depth of the employed methods. We will cover advances in scientific visualization of fields and particles, visual analytics of document collections and movement patterns as well as cognitive aspects.

Video Event Understanding with Pattern Theory

Sudeep Sarkar

Computer Science and Engineering, Associate Vice-President for Research &
Innovation, USF Research & Innovation, 3702 Spectrum Blvd., Suite 175,
Tampa, FL 33612-9444, USA
sarkar@usf.edu
http://www.usf.edu/engineering/cse/people/
sarkar-sudeep.aspx

Abstract. Successful automatic understanding of video content is essential for many computer vision-based applications. These applications generally focus on recognizing human interactions, which are typically understood as events that involve people performing actions with objects or other people. The main challenges are threefold. First, it is difficult to handle the enormous variety of interactions present in different instances of a complex event. The second challenge involves rejecting clutter of extraneous objects (detected in scenes) so that only the participating ones are included in the interpretation. The third challenge stems from low-level processing errors in segmentation, tracking, and classification of objects and actions. How we do leverage the success of deep-learning/machine learning based labeling methods into building a detailed semantic description of a complex event? There are many possible approaches; however, there is one approach that we have found to be particularly exciting. It is a combinatorial approach based on Grenander's pattern theory the classic version. In this talk I will share with you our approach and successes we have had in formulating a comprehensive and elegant formulation that is able to handle all these three kind of challenges, without the need for an extensive training dataset.

Cognitive Sciences: Tomorrow's Perspective

Manas Mandal

Life Sciences, DRDO, Ministry of Defence, Government of India, DRDO
Bhawan, Rajaji Marg, New Delhi 110105, India
mandalmanask@yahoo.com

Abstract. Cognitive science is a subject matter which is based on traditional methods in behavioral science with a focus on construct and theory development. Of late the focus in cognitive science has been changing with a change in operational definition of the construct of cognition that includes the components of neuroscience and computational science. From the perspective of behavioral science, this is an important development since many unanswered questions in cognitive science could be addressed, for example, how cognitive input in the brain is processed, why do we fail to process information under stress, Is it possible to bring autonomy in rudimentary human functions, can we develop decision support tool to reduce the burden of information overload, etc. The multidisciplinary nature of cognitive science today has helped bridge the gap between technology development and human skill development. Human cognitive threshold is thus being enhanced with the help of technology or being replaced with alternative / assistive technology. Efforts are also being made to reproduce the cognitive ability in man-made pre-programmed robot. To conduct such research, it is important to engage the disciplines of neuroscience, computer science and behavior science under the banner of cognitive science. The present talk will focus on cognitive science with an interdisciplinary perspective with an aim to propose some research programs that will help benefit developing technology to enhance, support, read or reproduce human capability.

Cognitive Sciences: Tomorrow's Perspective

Contents

HCI Applications and Technology

Interface and Systems

Intelligent Interfaces

Towards Learning to Handle Deviations Using User Preferences in a Human Robot Collaboration Scenario

Sharath Chandra Akkaladevi[1,2(✉)], Matthias Plasch[1],
Christian Eitzinger[1], Sriniwas Chowdhary Maddukuri[1],
and Bernhard Rinner[2]

[1] Profactor GmbH, Im Stadtgut A2, Steyr-Gleink 4407, Austria
sharath.akkaladevi@profactor.at
[2] Institute of Networked and Embedded Systems, Alpen-Adria-Universität
Klagenfurt, Klagenfurt, Austria

Abstract. In a human robot collaboration scenario, where robot and human coordinate and cooperate to achieve a common task, the system could encounter with deviations. We propose an approach based on Interactive Reinforcement Learning that learns to handle deviations with the help of user interaction. The interactions with the user can be used to form the preferences of the user and help the robotic system to handle the deviations accordingly. Each user might have a different solution for the same deviation in the assembly process. The approach exploits the problem solving skills of each user and learns different solutions for deviations that could occur in an assembly process. The experimental evaluations show the ability of the robotic system to handle deviations in an assembly process, while taking different user preferences into consideration. In this way, the robotic system could both benefit from interaction with users by learning to handle deviations and operate in a fashion that is preferred by the user.

Keywords: Learning to handle deviations · User preferences · Human robot collaboration · Ontology based knowledge representation

1 Introduction

Human robot collaboration (HRC) allows to combine the cognitive strength of humans together with the physical strength of robots and can lead to numerous applications [5]. For a close collaboration with humans, the robotic systems should attribute meaning to beliefs, goals and desires, collectively called "mental models" of humans during a particular task [11]. This would not only allow the robotic system to understand the actions and expressions of humans within an intentional or goal-directed architecture [14], but also for the human operators to better understand the capabilities of the robotic system [11]. In context of HRC, user preferences communicate the human's desire to achieve a goal (based on the belief that the goal is possible), either by themselves or in collaboration with the robotic system. User preferences can be communicated implicitly (via the actions done by the human) or explicitly (with the help offline knowledge representation). User preferences aid the robotic system in

A. Basu et al. (Eds.): IHCI 2016, LNCS 10127, pp. 3–14, 2017.
DOI: 10.1007/978-3-319-52503-7_1

creating and understanding the perspective of the human and collectively forms the basis for building the "mental model" of the human.

A detailed survey on human robot interaction, emerging fields and applications is given in [3, 6]. In this paper we present an HRC approach capable of handling deviations, where the robot interacts with the human by considering the 'preferences' of the human. By considering user preferences, the robotic system can take actions from the perspective (mental model) of the human to facilitate a 'natural' interaction. HRC should deal with the non-deterministic factor of the environment with human at its center, but it could also benefit from the cognitive and problem solving skills of the human. In recent works [7, 15, 18], it was shown that given the presence of the human, the robotic system can take advantage of human feedback and learn a given task in short time using Interactive Reinforcement Learning (IRL) techniques [15]. Inspired by this success, we propose an extension to the Interactive Reinforcement Learning (eIRL), which would allow the robotic system to take the aid of the human (asking which action to take next, receiving feedback) to learn to cope with deviations/novel situations as they occur. The advantage of allowing the user to teach the robot in coping with deviations is that each user has an own way of solving a given problem (communicated via user preferences). This allows the robot to learn different possibilities (when they exist) to deal with a deviation and use this knowledge to better collaborate and cooperate with the human. This also helps the robotic system to better personalize its behavior to a given user and their corresponding preferences.

In order to showcase the deviation handling capability of the proposed approach, an assembly process use case that deals with assembling objects into a box in collaboration with the human is chosen. This use case consists of four real-world objects, namely *heater*, *tray*, *ring* and *base* as shown in Fig. 1a. The goal of the use case is to assemble these objects by placing each of them inside a box. An example execution of pick and place operation of the *heater* object is shown in Fig. 1b. It is important to note that the manipulation possibility of each object either by the robot or the human is dependent on the construction (e.g., heavy, objects that require complex grasp - *base*) and configuration of the objects in the workspace.

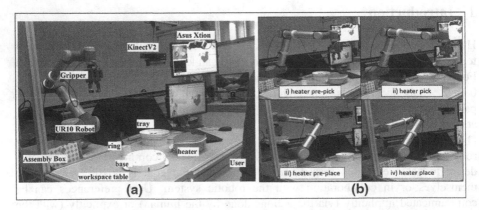

Fig. 1. (a) Use Case setup; (b) Pick and Place operation of the heater object

In this article, an HRC architecture that is capable of dealing with uncertainties and deviations of task executions using eIRL is presented. The architecture exploits the assistance of the human (who interacts with the robot to teach it) in the HRC loop, to learn and recover from deviations/novel situations that can occur during a task execution, using an interactive reinforcement learning algorithm. The approach where the robotic system:

- assesses the feasibility of executing the goal (to a certain level) before performing it
- **proactively suggests** possible course of action/s while considering the robot state (proprioception), the environment and the **preferences of the user**
- **enables technically unskilled users** to teach and customize the robot's behavior to their preferred manner
- **provides an interface** to communicate and learn from the human the necessary/alternative steps to take in case of deviations

The emphasis is on dealing with the deviations that occur in the action planning phase of the task. The low-level deviations that occur while carrying out the motion planning (path planning, collision avoidance) of the robot, while executing the task are not in the focus of this work.

The remaining part of the paper is organized as follows: In Sect. 2, a brief description of the existing state of the art approaches that deal with RL to learn a complete task are presented. Then in Sect. 3, the functional components and the architecture of the extended IRL is described. The theoretical background and the algorithmic description of eIRL is given in Sect. 4. Finally, the experimental setup and evaluation of the proposed eIRL is discussed in Sect. 5 followed by concluding remarks and possible future work described in Sect. 6.

2 Related Work

In order for the collaborative robots to work hand in hand with human operators the robotic system should be able to deal with complex and continuously changing environments. Thorough offline modeling of the environment and task conditions is not a feasible solution. Hence the robotic system should learn to respond to the environment by performing actions in order to reach a goal. The problem of agents learning in complex real-world environments is dealt with by numerous approaches [3, 4, 6]. Reinforcement learning (RL) [16] is one such popular approach that deals with learning from interaction, to teach the agent how to behave (which actions to perform when) in order to complete a task. The main aim in RL is to maximize a cumulative reward that is attained by taking some actions in the environment.

In RL, the agent learns over discrete time steps by interacting with its environment and gaining experience about the outcome. However, in order to reach an optimal policy (the set of actions that lead to the maximum reward), the RL approach requires substantial interaction with the environment. Depending on the nature of the task, RL results in a memory intensive storage of all state action pairs [8]. Another disadvantage of RL is its slow convergence towards a satisfactory solution. Despite such drawbacks, recent works on learning have shown considerable interest in using RL [7, 10, 15, 18].

The basic idea is to use the human teacher in the loop to provide feedback and hence speedup the convergence time in RL for reaching an optimal policy. In conventional RL, a reward is a positive or negative feedback for being in the current state (or for a particular state-action pair). Thomas et al. [18] introduced a run-time human feedback as a reward to the IRL approach and argue that this approach is beneficial for both the human teacher and the learning algorithm. Their study suggests that users (human teachers) employ the feedback as a single communication channel for various communicative intents-feedback, guidance, and motivation. Inspired by this approach, Suay et al. [15] study how IRL can be made more efficient for real-world robotic systems. In this approach, the user provides rewards for preceding actions of the robot and additionally provides guidance for subsequent actions. They further show that this guidance reduces the learning time of the robot and that it is more evident when large state-space size (number of interactions in the environment for the robot) is considered.

Knox et al. [9] propose a framework to train an agent manually via evaluative reinforcement, using real-valued feedback on its behavior from a human trainer. This allows the human trainer to interactively shape the agent's policy (interactive shaping) and thereby, directly modify the action selection (policy) mechanism of the IRL algorithm. Rather than in influencing the policy indirectly through a reward, Griffith et al. [7] use the user feedback in making a direct statement about the policy itself. Knox et al. [10] extended their previous work in [9] by applying the TAMER framework to real-world robotic system. Rozo et al. [12, 13] propose an approach for a human robot interactive task, where the robot learns both the desired path and the required amount of force to apply on an object during the interaction.

In this article, the following contributions are made to the state of the art:

- an extension to the Interactive Reinforcement learning algorithm (eIRL), that enables the robotic system to learn from the human, to deal with deviations.
- at each instant the robotic system encounters a deviation, it proactively suggests a list of actions possible by the robot in it's current state. This is achieved through efficient representation of the task knowledge using web ontology web language (OWL), see Sect. 3 for more details. The advantage Learning to Handle Deviations for Human Robot Collaboration 5 is that even untrained users can effortlessly teach the robot (how to deal with deviations) by choosing an appropriate action from the list provided.
- unlike the other approaches, the human feedback is divided into two: (a) feedback to choose an action policy from the list of options available (b) like/dislike feedback to function as a reward for an action executed by the robot, when a deviation is encountered.

3   eIRL Architecture

In order to collaborate and cooperate with the human in HRC scenario, the following components are developed within the extended IRL (eIRL) architecture as shown Fig. 2: (a) **perception and interpretation capabilities**: to understand and interpret the current state of the environment from sensor data (b) **knowledge representation**: to represent different aspects of the task carried out in the HRC environment that include representing the abilities and activities of the human, interplay between human activities and object configurations, robot's own self with respect to the task, etc. (c) **learning and reasoning**: the robotic system is equipped with learning capabilities that allows the accumulation of knowledge over time to enhance in particular its abilities to perceive the environment, to make decisions, to behave intelligently, and to interact naturally with humans. The robotic system also reasons about its current state to plan its own actions accordingly to aid in completing the task (d) **action planning**: to generate plans for future actions to achieve the given task (goals). This includes task planning, scheduling and observation planning, as well as planning under uncertainty for an efficient human robot collaboration. These plans generated need to be carried out in real world where the robot plans its path (path/navigation planning) and manipulates the environment accordingly. Note that the learning, reasoning and the action planning components form the heart of the eIRL architecture. (e) **functional component layer**: provides the necessary low-level sensory information like real-time 3D tracking of objects (implemented similar to [2]), current action of the human [1] together interfaces to robotic manipulation and a GUI-interface for human robot interaction.

Knowledge Representation in the proposed architecture is based on KnowRob [17]. The main reason for choosing KnowRob is that it provides the following knowledge processing features: (a) mechanisms and tools for action centric representation (b) automated acquisition of grounded concepts through observation and

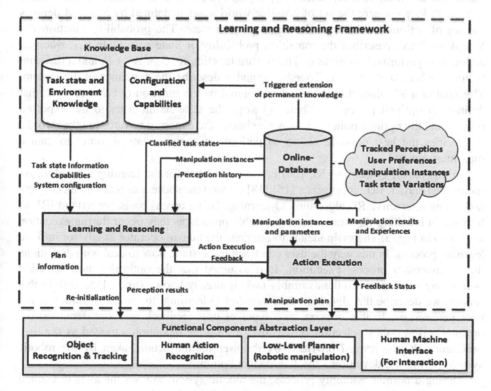

Fig. 2. Architecture diagram of the extended Interactive Reinforcement Learning approach

experience (c) reasoning about and managing uncertainty, and fast inference. The knowledge is represented using ontology (description logics) in the Web Ontology Language (OWL) and SWI Prolog is used for loading, accessing and querying the ontology. The representation consists of two levels: *Classes* that contain abstract terminological knowledge (type of objects, events and actions - taxonomic fashion) and *Instances* which represent the actual physical objects or the actions that are actually performed. The link between *Classes* and *Instances* is given in *Properties*, which defines if an *Agent* ∈ {*Human, Robot*} can perform a particular action (defined in *Classes*) on/with a *Target* ∈ {*Objects, Robot, Human*}. A Relation denoted by a triple <*Agent, Property, Target*> defines a particular aspect of the assembly task. For example, <*Robot, pick, ring*> conveys that the robot (is capable and) should pick the ring object. A detailed description of KnowRob and its features are given in [17].

4 eIRL Algorithm Description

Reinforcement Learning (RL) is an area of machine learning that defines a class of algorithms that enable a robot to learn from its experience. Reinforcement learning in our case is defined using the standard notation of the Markov Decision Process (MDP). In a MDP, any state s_{t+1} occupied by the robot is a function of its previous state s_t and the action taken at, in other words $s_{t+1} = f(s_t, a_t)$. A MDP is denoted by the 5-tuple <S, A, P, R, γ>, where the set of possible world states is defined by S, and A denotes the set of actions available to the agent in each state. The probability function P : $S \times A \rightarrow Pr[S]$, describes the transition probability of State s_t to State s_{t+1}, when an action a_t is performed on State s_t. The reward function $R : S \times A \rightarrow R$, and a discount factor γ, where $0 \leq \gamma \leq 1$. Together P and R describe the dynamics of the system. The goal of a RL algorithm is to find an approximation function $Q : S \times A \rightarrow R$, that defines an optimal policy π. Where, Q maps the state-action pairs to the expected reward and the optimal policy : $S \rightarrow A$ maximizes the expected reward. In other words, π specifies the best possible action to perform in a given state in order to gain a maximum reward.

In this paper, Q-learning RL [20] is used as a basis for the learning algorithm. As mentioned earlier, IRL approaches [15] [18] showed the success of learning a complete task using interactive RL algorithm, Q-learning. Using this as basis, we extend IRL to be used in handling deviations in an assembly process as they occur during execution as shown in Fig. 3. The main idea is to integrate the human operator as advisor into the learning process. In this way the user can teach the system, how to deal with deviations during assembly process execution. It is assumed that the optimal action selection policy, say π^*, to perform the assembly task, is already known (using [15, 18]). In this section we describe the algorithm that modifies Q-learning, to handle deviation using user preferences. In this context, two types of user feedback are considered (a) the optimal action policy selection by the user during the deviation - treated as the user preference (b) the reward provided by the user for the action taken by the robotic system during the deviation - the like/dislike option.

During a normal assembly process, the robotic system follows the already learned optimal policy π^* and performs actions a_{ti} accordingly based on its current state st to

1: Initialize Q table Q^* containing the action values $Q^*_{t,i}$, where $1 \leq t \leq |S|$ and $1 \leq i \leq |A_t|$ are the indices of states s_t and actions a_{ti} possible at the given state. Q^* leads to an optimal assembly process execution policy π^*, maximizing the reward

2: **for** each execution episode of the assembly process **do**
3:     Set assembly process to initial state $s_t \mid t = 1$
4:     **for** each execution step **do**
5:         Execute action a^*_{ti} in state s_t following policy π^*
6:         **if** no deviation occurred while executing a^*_{ti} **then**
7:             Observe resulting state s_{t+1}
8:             Q table Q^* is unchanged
9:         **else**
10:             Propose deviation handling actions $a^\Delta_{ti,j} \in A^\Delta_{ti} \mid j \in \mathbb{N}$
11:             **if** Action value $Q^*_{t,i} \not\sim \vec{Q}^\Delta_{ti}$ **then**
12:                 Initialize deviation Q vector $\vec{Q}^\Delta_{ti}$ with $Q^\Delta_{ti,j} = 0$
13:                 Associate $Q^*_{t,i}$ with $\vec{Q}^\Delta_{ti}$: $Q^*_{t,i} \sim \vec{Q}^\Delta_{ti}$
14:             **if** $\vec{Q}^\Delta_{ti}$ holds the initial values, i.e. $\max Q^\Delta_{ti,j} = 0$ **then**
15:                 User shall select a deviation handling action $a^\Delta_{ti,k} \mid 1 \leq k \leq |A^\Delta_{ti}|$
16:             **else**
17:                 Take handling action $a^\Delta_{ti,k} \leftarrow \arg \max\limits_{a^\Delta \in A^\Delta_{ti}} Q^\Delta_{ti,j}$
18:             Observe resulting state s_{t+1}, if $a^\Delta_{ti,k}$ was unsuccessful then go to ln.14
19:             Receive user feedback (*like/dislike*), determining the reward $r^\Delta_{t,k}$
20:             **for** each $Q^\Delta_{ti,j} \in \vec{Q}^\Delta_{ti}$ **do**

21:
$$Q^\Delta_{ti,j} \leftarrow Q^\Delta_{ti,j} + \alpha \left[r^\Delta_{t,j} + \gamma \max\limits_{1 \leq l \leq |A_{t+1}|} Q_{t+1,l} - Q^\Delta_{ti,j} \right] \mid \alpha = 1, \gamma = 0$$

22:                 If $j = k \implies a^\Delta_{ti,j} = a^\Delta_{ti,k}$ **then**
23:                     **if** *userFeedback = like* **then**
24:                         $r^\Delta_{t,j} \leftarrow r_{max}$
25:                     **else**
26:                         $r^\Delta_{t,j} \leftarrow 0$
27:                 **else**
28:                     $r^\Delta_{t,j} \leftarrow 0$
29:     $s_t \leftarrow s_{t+1}$

Comments
Ln 10: j is the index of possible deviation handling actions $a^\Delta_{ti,j}$
Ln 11: Each action value $Q^*_{t,i}$ can be associated with a deviation Q vector $\vec{Q}^\Delta_{ti}$
Ln 11: The symbol $\not\sim$ means "not associated with"
Ln 12: $Q^\Delta_{ti,j}$ is the action value for applying the deviation handling action $a^\Delta_{ti,j}$ to s_t, after encountering a deviation in action a^*_{ti}
Ln 18: It is assumed that the resulting state s_{t+1} is a known state of the assembly process
Ln 23: The reward $r_{max} \in \mathbb{R}^+$ is given, if the deviation handling result is *liked* by the user

Fig. 3. Algorithm to Handle deviations based on Q-Learning

progress to the next state s_{t+1}. As described earlier, the robotic system is enabled with state of the art action recognition and object tracking perception capabilities coupled with knowledge representation framework to comprehend the current status of the assembly task. After completing an action a_{ti} the robotic system observes the resulting state, if the resulting state is the expected next state s_{t+1}, the next optimal action policy in π^* is carried out and so on until the end. However, due to the dynamic nature of the environment, though an action is chosen by the robotic system, the task execution might not be successful and is called a deviation. A deviation could occur due to (a) the chosen action cannot be performed - e.g., object out of reach, object not available (b) expected resulting state s_{t+1} was not observed after the action a_{ti} is performed. A typical example could be an object was not graspable due to sensor error or the object configuration. It could also happen that the object might be out of reach of the robot. In such cases, it is not possible for the robotic system to continue with the assembly task as the learned optimal policy action cannot be completed.

Let us say action a_{ti} is an optimal policy action (given π^*) taken in state s_t and a deviation Δ_{ti} has occurred. In this case, the robotic system (depending on the current state, its capabilities, objects, knowledge base) proactively suggests a set of deviation handling actions $a_{ti,j}^{\Delta}$ possible in the current state s_t. Since deviations are special cases that occur during the assembly process, the original optimal policy should not be directly affected. Therefore, a deviation Q vector $\vec{Q}_{ti}^{\Delta}$ is associated with this deviation Δ_{ti} that occurred for this state-action pair $\{s_t, a_{ti}\}$ and its corresponding entry $Q_{t,i}^*$ in the Q table. If no vector $\vec{Q}_{ti}^{\Delta}$ existed previously for Δ_{ti}, then it is created and its values $Q_{ti,j}^{\Delta}$ are uniformly initialized to zero. This is generally the case, if this deviation was occurring for the first time. The robotic system then waits for the user to select an action policy ($a_{ti,j} \in A_{ti}^{\Delta}$) suitable for the deviation, from the list of suggested actions. In order to suggest the list of possible deviation handling actions A_{ti}^{Δ}, the robotic system observes the current status in the assembly process, available objects and robot's capabilities and then proposes a list of actions to progress the assembly process. Once the user selects an action policy, the action is carried out and the resulting state s_{tnext} is observed. Whenever, a deviation occurs the robotic system learns the optimal action policy in case of that deviation myopically i.e., $\gamma = 0$ [10]. In other words, it is assumed that no matter what deviation action policy is chosen by the user, it will progress the assembly process from the deviation to a known state s_{known}, from which an optimal state-action pair exists. This kind of learning from human feedback eliminates any exploration and results in a deviation handling policy that is only as good as the decision made by the user.

Upon successful completion of the suggested action policy ($s_{tnext} = s_{known}$), the robotic system asks for a *reward feedback*. The user at this stage can either *like* or *dislike* the robot's performance, as shown in Fig. 4. If the action is *liked* then the respective Q value for that action in the deviation $\vec{Q}_{ti}^{\Delta}$ vector is set to a maximum reward value and the other action entries are set to zero. In case of *dislike*, all the Q values in $\vec{Q}_{ti}^{\Delta}$ are set to 0. The execution is then continued with the optimal policy. If the deviation handling action was not successfully executed ($s_{tnext} \neq s_{known}$), the robotic system goes back and suggests possible deviation actions again. In this way the robotic system learns how to deal with the deviation that occurred for that state-action pair.

At a later point in time, if the robotic system encounters a deviation Δ_{ti}, there already exists a $\vec{Q}_{ti}^{\Delta}$ with initialized values. The robotic system then directly chooses the action policy $a_{ti,j}^{\Delta}$ in the $\vec{Q}_{ti}^{\Delta}$ that leads to a maximum reward value. After performing the deviation handling action, the robotic system asks the user for reward feedback. If the

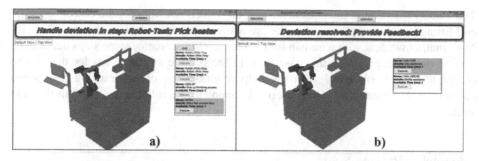

Fig. 4. Example of interaction cards provided while handling deviations in the assembly process (a) shows the list of deviation handling actions possible when robot encountered a deviation while Pick&Place heater action and (b) shows the like/dislike feedback interaction card presented to the user after successful handling of the deviation

user then *likes* the action performed by the robot, the Q values of the $\vec{Q}_{ti}^{\Delta}$ are updated accordingly and the assembly process proceeds as usual. Since the occurrence of these deviations are separated in time (a deviation can occur today and is repeated after an year), it is possible that user might like to teach the robot a different action policy $a_{ti,diff}^{\Delta}$ for handling that deviation Δ_{ti}. In such cases, the user *dislikes* the action performed by the robot which allows the robot to learn a new deviation action policy $a_{ti,diff}^{\Delta}$ for that deviation Δ_{ti}.

5 Experimental Setup

As shown in Fig. 1a, the experimental setup consists of a UR10 [19] robotic arm with 6 degrees of freedom, a SCHUNK 2 finger electric parallel gripper and two commercially available RGB-D sensors each equipped with object tracking and action recognition functionalities respectively. As mentioned in Sect. 3, the functional component layer with the help of sensors and actuators as shown in Fig. 1a provides the reasoning and learning module with necessary perception (object tracking and action recognition) and actuation (robot and gripper state combined to deduce if an action execution was successful) information.

The assembly process use case presented in this paper concerns itself with assembling the given objects (*heater, base, tray, ring*) in to a box. Depending on the capabilities of the *Agent* $\in$ {*Human, Robot*} and the object properties (configuration, construction) defined in the knowledge representation an optimal policy of a normal assembly process is defined as given in Fig. 5 and is assumed to be known. Figure 5 shows the optimal policy sequence (Step 1, Step 2, Step 3, Step 4 respectively) that expresses 'what' action is required to be performed by 'which' *Agent* and on 'which' *Target*. In this section, the ability of the system to learn, how to deal with deviations that occur during online execution (which were not encountered during training) of the assembly process is evaluated. During a normal assembly, Step 1 to 4 are executed in a sequence following the optimal policy and thereby gaining a maximum reward.

The *Pick&Place* action includes localizing the object, reaching for the object, grasping and lifting it and then placing it in the pre-assigned assembly box as shown in Fig. 1b. Note that, as we deal with a human robot collaboration scenario, some steps needs to be carried out by the user, in this case Step 1. The robotic system waits for the user to complete Step 1, observed with the help of embedded object tracking and action recognition functionalities. Once Step 1 is complete the robot proceeds with the next steps.

Assembly Process Step $< Agent, Property, Target >$	Possible Deviation Handling Actions
Step 1: User *Pick&Place base*	Note: only user can perform this action
Step 2: Robot *Pick&Place heater*	*retry*; *Pick&Place tray*; *Pick&Place ring*; *giveup*
Step 3: Robot *Pick&Place tray*	*retry*; *Pick&Place ring*; *giveup*
Step 4: Robot *Pick&Place ring*	*retry*; *giveup*

Fig. 5. Optimal action policy of the assembly process and the respective deviation handling actions available at each step

A deviation can occur at any step in the assembly process and the possible deviation handling actions at each step are given in Fig. 5. These deviation handling actions are presented as interaction cards to the user as shown in Fig. 4, who can then select an action by simply clicking on the 'Execute' button. The deviation handling action *retry* entails that the same action step has to be retried, e.g., if in Step 2 a deviation occurs and the user selects *retry*, then Step 2 will be executed again. The deviation handling action *give-up* communicates to the robot that assembly process should be stopped and the robot should move to an initialized position. A total of 25 executions of the assembly process by 5 users (roboticists) with 5 executions per user were carried out. Figure 6 shows five different profiles created during online execution. Each profile has three entries, which consists the chosen deviation handling action by that user for Step 2, 3 and 4 respectively. During the online execution of the 5 assembly process per user, different deviations possible were introduced at random such that, overall the user teaches a deviation handling action to the robot for each step of the assembly process.

Learning to handle deviation in this fashion has the following advantages: (a) the robot is not required to explore all possible deviation handling actions for each step to find a solution. This reduces the amount of time exponentially as shown in Fig. 6 depending on the number of deviation handling actions possible at a given step where a deviation occurred (b) the robotic system is capable of handling novel situations (situation not previously encountered) as they occur online during the assembly process with the help of the user interaction. Also, each user has their own profile which will enable the robot to interact with different users according to their chosen fashion. In the case of a new user, the robot learns deviation handling actions specific to that user and hence collaborates more closely using the user specific preference.

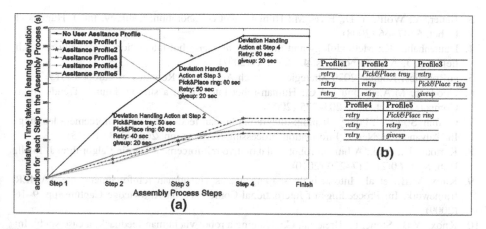

Fig. 6. (a) Shows the time taken for the robot to handle deviations in presence and absence of the user assitance. (b) Deviation Handling Profiles of 5 users, learned over a course of 5 assembly process executions per user (total 25 Executions). The three entries in each profile column describe the deviation handling action chosen by that user for the assembly process steps (as given in Fig. 5) 2, 3 and 4 respectively.

6   Conclusion and Future Work

In this paper we have extended the IRL to enable the robotic system to handle deviations in an assembly process that occur during real-time execution. The eIRL exploits the presence of the human user in the human robot collaboration scenario by interacting with the user for assistance and feedback. The robotic system proposes a set of possible solutions to the user, given a deviation at a particular step in the assembly process. The proposed set of solutions is derived from the knowledge of the assembly process, already known optimal policy, robot capabilities and the set of objects currently present in the workspace. The user assistance is divided into a direct deviation action policy choice and a reward feedback. This enables the system to keep track of user preferences and interact with them more intuitively. An interesting future work would be to combine the already existing preferences of users (the choice of the existing users) and use it to interact with a new untrained user. Also, the evaluation was carried out with roboticists and we would like to extend this evaluation by including users with varying levels of experience with robots.

Acknowledgment. This research is funded by the projects KoMoProd (Austrian Ministry for Transport, Innovation and Technology), and CompleteMe (FFG, 849441).

References

1. Akkaladevi, S.C., Heindl, C.: Action recognition for human robot interaction in industrial applications. In: CGVIS 2015, pp. 94–99. IEEE (2015)
2. Akkaladevi, S., et al.: Tracking multiple rigid symmetric and non-symmetric objects in real-time using depth data. In: ICRA 2016, pp. 5644–5649 (2016)

3. Bauer, A., Wollherr, D., Buss, M.: Human-robot collaboration: a survey. Int. J. Humanoid Robot. **5**, 47–66 (2008)
4. Dautenhahn, K.: Methodology and themes of human-robot interaction: a growing research field. Int. J. Adv. Robot. Sys. **4**, 103–108 (2007)
5. euRobotics: Robitcs 2020 Strategic Research Agenda for Robotics in Europe (2013)
6. Goodrich, M.A., Schultz, A.C.: Human-robot interaction: a survey. Found. Trends Hum. Comput. Interact. **1**(3), 203–275 (2007)
7. Griffith, S., et al.: Policy shaping: integrating human feedback with reinforcement learning. In: Advances in Neural Information Processing Systems, pp. 2625–2633 (2013)
8. Kartoun, U., et al.: A human-robot collaborative reinforcement learning algorithm. J. Intell. Rob. Syst. **60**(2), 217–239 (2010)
9. Knox, W.B., et al.: Interactively shaping agents via human reinforcement: the TAMER framework. In: Proceedings of International Conference on Knowledge Capture, pp. 9–16 (2009)
10. Knox, W.B., Stone, P., Breazeal, C.: Training a robot via human feedback: a case study. In: Herrmann, G., Pearson, M.J., Lenz, A., Bremner, P., Spiers, A., Leonards, U. (eds.) ICSR 2013. LNCS (LNAI), vol. 8239, pp. 460–470. Springer, Heidelberg (2013). doi:10.1007/978-3-319-02675-6_46
11. Lee, S., et al.: Human mental models of humanoid robots. In: IEEE International Conference on Robotics and Automation, pp. 2767–2772 (2005)
12. Rozo, L., et al.: Learning collaborative impedance-based robot behaviors. In: Twenty-Seventh AAAI Conference on Artificial Intelligence, pp. 1422–1428 (2013)
13. Rozo, L., et al.: Learning force and position constraints in human-robot cooperative transportation. In: Robot and Human Interactive Communication, pp. 619–624 (2014)
14. Scassellati, B.: Theory of mind for a humanoid robot. Autonomous Robots **12**(1), 13–24 (2002)
15. Suay, H.B., Chernova, S.: Effect of human guidance and state space size on interactive reinforcement learning. In: 2011 Ro-Man, pp. 1–6. IEEE (2011)
16. Sutton, R.S., Barto, A.G.: Reinforcement learning: an introduction, vol. 1, no. 1. MIT Press, Cambridge (1998)
17. Tenorth, M., et al.: KnowRob: a knowledge processing infrastructure for cognition enabled robots. Int. J. Robot. Res. **32**, 566–590 (2013)
18. Thomaz, A.L., et al.: Reinforcement learning with human teachers: understanding how people want to teach robots. In: ROMAN 2006, pp. 352–357. IEEE (2006)
19. Universal Robots, UR10 robot - a collaborative industrial robot. http://www.universal-robots.com/products/ur10-robot/. Accessed 09 Aug 2016
20. Watkins, C.J., Dayan, P.: Q-learning. Mach. Learn. **8**(3–4), 279–292 (1992)

Position Based Visual Control of the Hovering Quadcopter

Atulya Shivam Shree, Radhe Shyam Sharma,
Laxmidhar Behera$^{(\boxtimes)}$, and K.S. Venkatesh

Indian Institute of Technology Kanpur, Kanpur 208016, UP, India
{atulya,sharmars,lbehera,venkats}@iitk.ac.in

Abstract. Autonomous navigation of quadcopters in unstructured indoor environments is a major problem due to the difficulty of reliable position sensing. While outdoor applications can use GPS for reliable localization, working indoors will require the use of either laser range finders or some other sensors. If the indoor scene is unknown to a robot, the task of mapping new areas also becomes a necessity. The two processes are combined and run together in a framework of Simultaneous Localization and Mapping (SLAM). Our work is focused on using onboard cameras for the task of SLAM in an indoor scenario. Vision based techniques that do not use time of flight methods like laser range finders, have the potential to provide a low cost alternative framework for navigation. In this work, localization using a monocular SLAM framework on an unknown and unstructured scene, a cascaded position controller along with a Luenberger observer which can combine the data of Inertial sensors and vision based position to generate a complete velocity feedback for the system have been used. Sensor data fusion using EKF (Extended Kalman Filter) have been performed for scale estimation. The localization algorithm has been implemented on a quadcopter. Finally hovering experiment has been performed in an indoor lab based environment.

Keywords: Quadcopter · Autonomous · Navigation · Vision · VSLAM

1 Introduction

In recent years, quadcopters have gained a lot of popularity. There are numerous commercial products already developed and being sold in the market. The highly agile dynamics of a quadcopter allow it to easily takeoff and fly in any indoor or outdoor scenario. People are using them commercially for photography and for recording videos by mounting a camera on them. One of the major challenges in the field of micro aerial vehicles is to make them completely autonomous: this requires localization. In an outdoor environment it is possible to use GPS based position feedback to fly over a particular trajectory. But GPS does not work indoors and so while flying indoors, if no correction is provided by a manual operator the quadcopter will drift in any random direction due to the absence

© Springer International Publishing AG 2017
A. Basu et al. (Eds.): IHCI 2016, LNCS 10127, pp. 15–26, 2017.
DOI: 10.1007/978-3-319-52503-7_2

of precise location information. This calls for using vision or lasers for localizing the robot in an indoor environment.

Passive systems like VICON or active systems system like phoenix are very popular choices for lab based environments. It has been shown in [1] that even highly complex maneuvers and trajectory tracking can be done provided we have a highly accurate position feedback system. However system relies on external tracking hardware and is therefore limited to a lab environment.

For complete autonomy in indoor operations, it is necessary to implement SLAM techniques onboard the system. This can also allow it to be able to navigate through initially unknown terrains. Highly accurate SLAM with dense mapping of the scene has been demonstrated in [2] where the authors have used laser along with vision to generate a pointcloud of the environment. While laser range finders provide highly accurate depth data they are generally heavy and expensive. Hence there is a need for developing the SLAM framework by using cameras as the primary sensors without using laser range finders. The research on vision based SLAM is particularly important because it is believed that this is how we humans explore our surroundings. Vision based sensors are now extremely cheap, and hence developing a good algorithm for state estimation can make a product commercially viable to the general public.

In order to stabilize a robot over any particular scene two methods have been used IBVS (Image Based Visual Servoing) and PBVS (Position Based Visual Servoing). IBVS has been used in [3], and it uses spherical image based error in between the desired image and current image to directly generate control signals. However, it requires knowledge of the desired scene beforehand and hence is restricted to use with pre-defined markers and visual patterns. PBVS on the other hand generates the current position feedback of the camera by tracking certain inherent salient features in the scene. This method continuously generates a sparse feature point map of the surroundings and simultaneously performs localization with respect to it. To implement complete vision based SLAM, Parallel Tracking and Mapping (PTAM) [4] which uses a single camera has been the most popular choice. In the work in [5] the authors have used an AR Drone platform for sending the camera stream wirelessly over a WiFi network and implementing PTAM on an offboard laptop computer. In spite of delays in transmission the authors have been able to successfully stabilize the Micro Aerial Vehicle (MAV). The major disadvantage of using monocular camera based algorithms is that the pose measurements are scaled with respect to the metric measurements. To solve this problem work has been done on using stereo SLAM for application to MAVs. A hybrid approach using a slow stereo camera set along with fast PTAM based odometry has been presented in [6]. While PTAM uses only one camera, with a stereo configuration it is also possible to obtain accurate scale corrections and depth map estimation. In [7] the authors use a bottom facing camera for velocity estimation using optical flow and a forward facing stereo configuration for performing complete mapping and exploration. In [8] two sets of stereo cameras have been used, in the front and the bottom direction to provide complete state metric localization and mapping. Another technique for scale correction is to fuse the readings of vision with that of Inertial sensors. In [9] the readings of vision based monocular sensor have been combined with the readings of an IMU and Barometer to provide complete state feedback to the system.

In more advanced work [10–12] the authors have demonstrated complete state estimation by fusing the sensor readings of vision and IMU. The method does not use any other sensor and can provide metric state feedback by estimating the scale of the monocular SLAM system. However the system is very sensitive to initial values of scale and can fail to converge if the initial estimate is outside a certain bound. More recent work developed in 2014 include SVO (Semi-direct Visual Odometry) [13] which is a highly efficient and fast open source monocular SLAM algorithm. This algorithm has the ability to work at almost 60 fps on a computationally restricted onboard computer and even faster on consumer grade laptops. It achieves this by directly operating on pixel intensities for matching patches within different frames. This reduces the computational burden of extracting feature points at every new frame making the visual system more robust. This paper is mainly concerned with implementation of a vision based hovering framework on a quadcopter. A Luenberger observer has been designed which predicts the velocity using the accelerometer and the position feedback. For localization, monocular SLAM algorithm SVO has been used which is an open source package released for Robot Operating System (ROS) by [13]. This is followed by experimental demonstration of hovering over particular waypoints in 3D space.

The paper is organized as follows: sensors, observer & EKF are described in Sect. 2. Experimental results are provided in Sect. 3. Section 4 concludes the paper.

2 Localization and Position Control in Unstructured Environment

2.1 Hardware Description

Nayan quadcopter is used to perform experiment which is shown in Fig. 1. Description of its components are mentioned in Table 1.

Table 1. Description of Nayan hardware platform

Component	Description
Flight controller	ARM Cortex M4 32 Bit 168 MHz CPU
Onboard computer	Odroid
Camera	CMOS, Bluefox, 752 × 480
Ultrasonic sensors	Px4Flow
Battery	LiPo, 5200 mAh
Propeller	11″ × 4.5″
Weight	1.9 Kg

Fig. 1. Nayan quadcopter

2.2 Sensor Description

Inertial Measurement Unit. The IMU is the most crucial element of the quadrotor and is used by the attitude controller for maintaining a desired orientation. It consists of a gyroscope which measures the angular velocity in the body frame and an accelerometer which measures acceleration in the body frame. The Rotation matrix at any time is obtained by fusing the data of accelerometer with that of the gyro. Ideally a gyro measures the angular velocity of the system and is sufficient to estimate the orientation angles ϕ, θ, ψ, given the initial estimate. However there is a small bias in the angular velocity measurements which leads to a drift in the attitude, thus calling the need for fusion. More details on the working of Inertial navigation systems can be found at [14]. For this experiment it is assumed that the attitude fusion has already been done in the LLP (Low Level Processor) and we have been provided with the roll, pitch and yaw angles in High Level Processor (HLP) where the custom user code runs. This is utilized for obtaining the acceleration in the NED-b frame from the body frame. The data of accelerometer is governed by the following equation:

$$z = R_g^b(a_I - g) + b + n_a \tag{1}$$

where, z is the data measured by the accelerometer, R_g^b is the rotation matrix from Inertial to Body frame, a_I is the acceleration in the inertial frame, b is a bias in the accelerometer, g is the acceleration of gravity, n_a is noise in the sensor readings. A very important assumption taken to simplify all sensor fusion steps is that the yaw angle has been considered to be fixed. Under this assumption it is necessary that the position feedback to the observer is the NED-body fixed frame. This also leads to a simplified R_g^b matrix as the yaw term is not considered. Hence after these assumptions the final equation for acceleration in the NED-B frame is:

$$a_{nedb} = \begin{bmatrix} \cos\theta & \sin\theta\sin\phi & \cos\phi\sin\theta \\ 0 & \cos\phi & -\sin\phi \\ -\sin\theta & \sin\phi\cos\theta & \cos\phi\cos\theta \end{bmatrix} \begin{bmatrix} z_x \\ z_y \\ z_z \end{bmatrix} + \begin{bmatrix} 0 \\ 0 \\ 9.81 \end{bmatrix} \tag{2}$$

where, ϕ & θ are roll and pitch angle respectively. Note however that the above acceleration is not bias free and needs to be corrected before being further used.

Ultrasonic Distance Sensor. An ultrasonic sensor calculates the distance to an object by using time of flight data for a pulsed echo signal. In this experiment a Px4Flow board has been used comprising of an inbuilt ultrasonic sensor which provides direct metric measurements of the height of MAV from the ground. The data is sampled at a frequency of 10 Hz from the sensor.

2.3   Luenberger Observer

The output from the ultrasonic sensors is a depth reading available at 10 Hz. The VSLAM algorithm will also be able to give position coordinates at upto 25 Hz. The two are combined on the main computer and finally a metric position update is provided to the position controller running on the HLP of the flight controller. We also have the acceleration data coming in from the onboard IMU at a high frequency. With these available sensor data a luenberger observer has been designed similar to that used in [9,15]. Each of the x, y, z coordinates are assumed to be an independent system with an order 3 state for each.

$$\begin{bmatrix} \dot{x} \\ \ddot{x} \\ \dot{b}_x \end{bmatrix} = \begin{bmatrix} 0 & 1 & 0 \\ 0 & 0 & 1 \\ 0 & 0 & 0 \end{bmatrix} \begin{bmatrix} x \\ \dot{x} \\ b_x \end{bmatrix} + \begin{bmatrix} 0 \\ 1 \\ 0 \end{bmatrix} a_{nedb_x} \tag{3}$$

$$Y = \begin{bmatrix} 1 & 0 & 0 \end{bmatrix} \begin{bmatrix} x \\ \dot{x} \\ b_x \end{bmatrix} \tag{4}$$

Here, a_{nedb_x} is the acceleration in x axis which is taken as an input to this system, while $(x, \dot{x}, b_x)$ are the states. Y is taken as the system output which in our case is available to the flight controller from the main computer. The states can similarly be defined for x, y and z. a_{nedb} is obtained from Eq. 2. Based on the above formulation a Luenberger observer is designed. For a system defined as:

$$\dot{x} = Ax + Bu, \qquad Y = Cx$$

we have an observer,

$$\dot{\hat{x}} = A\hat{x} + Bu + L(y - \hat{y}) \tag{5}$$

The error $e = x - \hat{x}$ converges to zero provided the eigen values of the matrix $A - LC$ are all negative. Hence the values of the observer L are chosen appropriately.

2.4   Kalman Filter for Metric Position Feedback

Position Feedback from SVO. The visual subsystem can be thought of as a black box unit which ultimately gives the current pose as the feedback. In this experiment, only the position data has been used for controls while the IMU which is more reliable for roll and pitch angles, is used for the attitudes.

$$\vec{r}_v = \frac{1}{\lambda}(\vec{r} - \vec{r}_0) \tag{6}$$

where, $\vec{r}_v$ is the feedback coming from SVO, $\vec{r}$ is the actual 3D coordinate, r_0 is position at which the visual system was initialized, λ is some unknown scale factor. This output is with respect to an inertial frame which is aligned with the ENU-body frame at the initial point. At any time after initialization four states from the VSLAM are taken $(x_{vi}, y_{vi}, z_{vi}, \psi_{vi})$. These are first transformed to get the position in the gravity aligned body fixed frame. In the final step the position is transformed from the ENU (East North Up) coordinate system of SVO to NED (North East Down) for the NED-Body-fixed frame in which the control algorithm primarily works.

$$\vec{r}_{vb} = R_{vi}^{vb}\vec{r}_{vi}, \quad R_{vi}^{vb} = R_z(\psi) = \begin{bmatrix} \cos\psi & \sin\psi & 0 \\ -\sin\psi & \cos\psi & 0 \\ 0 & 0 & 1 \end{bmatrix}$$

where, ψ is the yaw angle.

EKF Framework for Estimation of Static States. Since there are multiple sensory inputs for the height of the quadcopter, a fusion step using EKF is performed so as to estimate the static states λ, z_0. The different equations are:

$$z_v = \left(\frac{z - z_0}{\lambda}\right) + n_v, \quad z_u = z + n_u, \quad \ddot{z} = a_{nedbz} + b + n_a \tag{7}$$

where, z_v is the depth from vision sensors, z_u is the depth from ultrasonic sensors, z is the position of the quadcopter on z axis. λ, z_0 are the static states for the VSLAM system, b is the static bias of the IMU. n_v, n_a, n_u are Gaussian noise with 0 mean and fixed variance. The state of the system at instant k is defined as

$$x_{k|k} = \begin{bmatrix} z & \dot{z} & b & \lambda & z_0 \end{bmatrix}^T \tag{8}$$

Prediction

$$x_{k+1|k} = F_k x_{k|k} + Bu, \quad P_{k+1|k} = F_k P_{k|k} F_k^T + V_k$$

where,

$$F_k = \begin{bmatrix} 1 & \Delta t & 0 & 0 & 0 \\ 0 & 1 & \Delta t & 0 & 0 \\ 0 & 0 & 1 & 0 & 0 \\ 0 & 0 & 0 & 1 & 0 \\ 0 & 0 & 0 & 0 & 1 \end{bmatrix}, \quad B_k = \begin{bmatrix} 0 & \Delta t & 0 & 0 & 0 \end{bmatrix}^T$$

Update for VSLAM

$$\hat{z}_v = \frac{1}{x_{k+1|k}(3)}(x_{k+1|k}(0) - x_{k+1|k}(4)), \quad v = z_v - \hat{z}_v$$

$$H_{v,k+1} = \begin{bmatrix} \frac{1}{x_{k+1|k}(3)} \\ 0 \\ 0 \\ -\frac{(x_{k+1|k}(0)-x_{k+1|k}(4))}{x_{k+1|k}^2(3)} \\ \frac{1}{x_{k+1|k}(3)} \end{bmatrix}^T$$

$$S = H_{u,k+1}P_{k+1|k}H_{u,k+1}^T + W_u, \quad R = P_{k+1|k}H_{u,k+1}S^{-1}$$

$$x_{k+1|k+1} = x_{k+1|k} + Rv, \qquad P_{k+1|k+1} = P_{k+1|k} - RH_{u,k+1}P_{k+1|k}$$

Update for Ultrasonic Sensor

$$v = z_u - x_{k+1|k}$$

$$H_{u,k+1} = \begin{bmatrix} 1 & 0 & 0 & 0 & 0 \end{bmatrix}$$

$$S = H_{u,k+1}P_{k+1|k}H_{u,k+1}^T + W_u, \quad R = P_{k+1|k}H_{u,k+1}S^{-1}$$

$$x_{k+1|k+1} = x_{k+1|k} + Rv, \qquad P_{k+1|k+1} = P_{k+1|k} - RH_{u,k+1}P_{k+1|k}$$

where,

$$P_{0|0} = \begin{bmatrix} 2 & 0 & 0 & 0 & 0 \\ 0 & 1 & 0 & 0 & 0 \\ 0 & 0 & 0.30 & 0 & 0 \\ 0 & 0 & 0 & 4 & 0 \\ 0 & 0 & 0 & 0 & 0.4 \end{bmatrix}, \quad V_k = \begin{bmatrix} 0.3^2 & 0 & 0 & 0 & 0 \\ 0 & 0.6^2 & 0 & 0 & 0 \\ 0 & 0 & 0.04^2 & 0 & 0 \\ 0 & 0 & 0 & 0.003\Delta t & 0 \\ 0 & 0 & 0 & 0 & 0.001\Delta t \end{bmatrix} \Delta t$$

here, standard notations are used for EKF. Symbol hat is used to represent the estimated parameter. W_u is taken as 0.02, W_v is taken as 0.1, Δt is taken as 0.02 s which corresponds to the fact that the IMU update works at 50 Hz on the system.

2.5 Velocity Feedback Based Controller

This controller assumes that the position feedback it receives is in the NED-body frame. The controller used in this experiment is a cascaded PID controller. It comprises of a velocity feedback loop and a position feedback. The velocity feedback loop obtains the current velocity using the observer designed in Eq. 5. The position feedback runs at a frequency of 50 Hz and it mainly generates the desired velocity command for the inner velocity control loop.

$$e_x = x_{ref} - \hat{x}, \quad e_{intgx} = \int e_x dt$$

$$v_{xref} = K_{px}e_x - K_{dx}v_x + K_{ix}e_{intgx}$$

$$e_{vx} = v_{xref} - \hat{v}_x, \quad e_{intgvx} = \int e_{vx}dt$$

$$a_{xref} = K_{pvx}e_{vx} - K_{dvx}a_{nedbx} + K_{ivx}e_{intgvx}$$

where, e_x is the position error in x, e_{vx} is the velocity error in x. In subscript ref stands for reference, whereas hat is used to denote estimated parameter. a_{xref} is the acceleration in x. The final output of this controller is a desired a_{xref}. a_{yref} and a_{zref} are obtained in a similar fashion. The acceleration now needs to be converted to a desired attitude angle and thrust. The small angle assumption is taken

$$\ddot{x} = -g\theta, \quad \ddot{y} = g\phi, \quad \ddot{z} = g - \frac{U_1}{m}$$

where U_1 is the thrust. Hence the equations for desired attitude angles can now be computed according to the inverse of the above relation.

$$\phi_d = \frac{a_{yref}}{g}, \quad \theta_d = -\frac{a_{xref}}{g}, \quad T_d = T_0 - a_{zref}$$

where

$$U_1 = b_{tc}(\Omega_1^2 + \Omega_2^2 + \Omega_3^2 + \Omega_4^2)$$

b_{tc} is the thrust coefficient, Ω denotes the angular velocity of the respective rotor. $\ddot{x}$, $\ddot{y}$ and $\ddot{z}$ denote the accelerations in x, y and in z direction respectively. m is the mass of the system. T_d and T_0 represent the desired thrust and hovering throttle respectively.

3   Experimental Results

In Fig. 2(a), we have included a picture of a real time experiment in which the quadcopter, safety net & pattern are marked. For experimenting indoors a pattern was placed on the ground. The pattern is used to localize quadcopter in a precise and robust manner using SVO. Features tracking using SVO is shown in Fig. 2(b). Natural features may or may not contain enough number of strong point features, which can potentially lead to tracking failure. That is why we have chosen an artificial pattern that is sufficiently rich in strong point features. A safety net exists around the operational area to keep other people safe in case of an accident. The entire code for this experiment runs on the onboard systems. We have chosen the values of Kalman filter parameters heuristically. Visual SLAM and Kalman state estimation run on the onboard computer whereas the Luenberger observer and the position controller run on the onboard HLP. The mavros driver is used to perform all required two way communication between the main onboard computer and the flight controller. In order to initiate the different ROS nodes running on the onboard computer, the user remote logs into the system via a WiFi network. The different nodes are then executed using ROS. There are five nodes running on the main onboard computer while one node runs on the desktop computer and is used for viewing the debug output of the VSLAM (Visual SLAM). The complete workflow and details of various nodes on ROS network are given in Fig. 3 and Table 2 respectively.

Fig. 2. (a) Indoor lab setup with hovering experiment in autonomous mode. (b) Feature tracking using SVO.

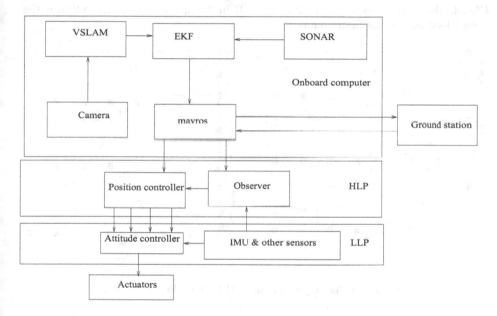

Fig. 3. System overview & complete work flow.

Table 2. Various nodes on the ROS network.

Node	Purpose
Mavros	For communicating with the onboard flight controller
Cam_driver	For extracting and publishing camera data
Px4flow	For extracting and publishing data acquired from the px4flow sensor
SVO	For running the VSLAM algorithm
Sensor_fusion	Fusing depth data from visual and ultrasonic sensors
Image_view	For viewing the debug output of the visual SLAM

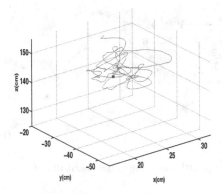

Fig. 4. Hovering at fixed point $(23, -37, 144)$ in 3D space, blue point indicates the target location. (Color figure online)

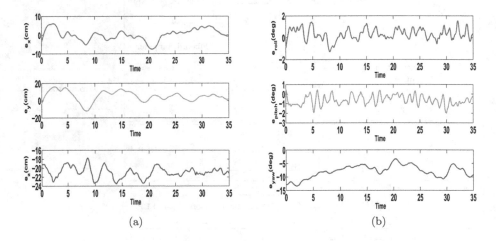

Fig. 5. (a) Position error plots. (b) Attitude error plots.

We used Bluefox (CMOS, 752×480) camera in our experiment. Localization using SVO is performed after camera calibration. For an experiment, the manual operator performs the takeoff and the landing procedures. Even while in autonomous flight, the operator must remain alert to take back control in case of any random behavior or loss of track. Hovering at the fixed point is shown in Fig. 4 while position and attitude error are given in Fig. 5(a) and (b) respectively. Error plots in Fig. 5(a) and (b) are taken from the controller inputs and outputs respectively. The central objective to obtain autonomous hovering without the help of any external positioning system like VICON. The video of the experiment can be seen in https://youtu.be/L7imtmpw8mU.

4   Conclusion

In this paper, vision based localization and mapping techniques have been explored with application to a micro aerial vehicle. The SLAM based technique has been found to perform good localization in unknown scenes. We have tested this framework in indoor environment using vision based feedback to achieve a hovering task. This work is useful to achieve waypoint navigation. Experimental results have confirmed the potential of this framework.

References

1. Mellinger, D., Shomin, M., Kumar, V.: Control of quadrotors for robust perching and landing. In: International Powered Lift Conference, pp. 205–225 (2010)
2. Shen, S., Mulgaonkar, Y., Michael, N., Kumar, V.: Multi-sensor fusion for robust autonomous flight in indoor and outdoor environments with a rotorcraft MAV. In: 2014 IEEE International Conference on Robotics and Automation (ICRA), pp. 4974–4981 (2014)
3. Hamel, T., Mahony, R.: Visual servoing of an under-actuated dynamic rigid-body system: an image-based approach. IEEE Trans. Robot. Autom. 18(2), 187–198 (2002)
4. Klein, G., Murray, D.: Parallel tracking and mapping for small AR workspaces. In: 6th IEEE and ACM International Symposium on Mixed and Augmented Reality, pp. 225–234 (2007)
5. Engel, J., Sturm, R., Cremers, D.: Accurate figure flying with a quadrocopter using onboard visual and inertial sensing. IMU 320, 240 (2012)
6. Schauwecker, K., Ke, N.R., Scherer, S.A., Zell, A.: Markerless visual control of a quad-rotor micro aerial vehicle by means of on-board stereo processing. In: Levi, P., Zweigle, O., Häußermann, K., Eckstein, B. (eds.) Autonomous Mobile Systems 2012. Informatik aktuell, pp. 11–20. Springer, Berlin (2012)
7. Fraundorfer, F., Heng, L., Honegger, D., Lee, G.H., Meier, L., Tanskanen, P., Pollefeys, M.: Vision-based autonomous mapping and exploration using a quadrotor mav. In: 2012 IEEE/RSJ International Conference on Intelligent Robots and Systems (IROS), pp. 4557–4564 (2012)
8. Yang, S., Scherer, S., Zell, A., et al.: Visual SLAM for autonomous MAVs with dual cameras. In: 2014 IEEE International Conference on Robotics and Automation (ICRA), pp. 5227–5232 (2014)
9. Achtelik, M., Achtelik, M., Weiss, S., Siegwart, R.: Onboard IMU and monocular vision based control for MAVs in unknown in-and outdoor environments. In: 2011 IEEE International Conference on Robotics and Automation (ICRA), pp. 3056–3063 (2011)
10. Nützi, G., Weiss, S., Scaramuzza, D., Siegwart, R.: Fusion of IMU and vision for absolute scale estimation in monocular SLAM. J. Intell. Robot. Syst. 61(1–4), 287–299 (2011)
11. Weiss, S., Achtelik, M.W., Lynen, S., Chli, M., Siegwart, R.: Real-time onboard visual-inertial state estimation and self-calibration of MAVs in unknown environments. In: 2012 IEEE International Conference on Robotics and Automation (ICRA), pp. 957–964 (2012)

12. Weiss, S., Siegwart, R.: Real-time metric state estimation for modular vision-inertial systems. In: 2011 IEEE International Conference on Robotics and Automation (ICRA), pp. 4531–4537 (2011)
13. Forster, C., Pizzoli, M., Scaramuzza, D.: SVO: fast semi-direct monocular visual odometry. In: 2014 IEEE International Conference on Robotics and Automation (ICRA), pp. 15–22 (2014)
14. Woodman, O.J.: An introduction to inertial navigation. University of Cambridge, Computer Laboratory, Technical report. UCAMCL-TR-696, vol. 14, p. 15 (2007)
15. Shree, A.S.: Vision based navigation of a quadcopter using a single camera. Master's thesis, Indian Institute of Technology, Kanpur, Kanpur, India (2015)

SG-PASS: A Safe Graphical Password Scheme to Resist Shoulder Surfing and Spyware Attack

Suryakanta Panda[✉] and Samrat Mondal[✉]

Department of Computer Science and Engineering,
Indian Institute of Technology Patna, Patna, India
{suryakanta.pcs15,samrat}@iitp.ac.in

Abstract. In general, it is difficult to remember a strong password i.e. a long and random password. So, the common tendency of a user is to select a weak alphanumeric password that is easy to remember. But the password which is easy to remember is also easy to predict. In contrast, the password that is very difficult to predict or requires more computation to break is also difficult to remember. To overcome this limitation of creating secure and memorable passwords, researchers have developed graphical password scheme which takes images as passwords rather than alphanumeric characters. But graphical password schemes are vulnerable to shoulder-surfing attack where an attacker can capture a password by direct observation. In this paper a graphical password scheme, namely SG-PASS is proposed which can prevent the shoulder-surfing attack by a human observer and also spyware attack, using a challenge response method.

Keywords: Graphical passwords · User authentication · Password security

1 Introduction

Authentication is an essential security component for many internet applications. It is a must for every system that provides secure access to confidential information and services. All the authentication techniques rely on at least one of the following methods:

- *Something you have:* it is also called as token based authentication. In this mechanism, each user has some physical identification objects which identify the user uniquely e.g. smart cards.
- *Something you know:* it is also called as knowledge based authentication. In this mechanism, a user has to remember an alphanumeric password or images for a graphical password.
- *Something you are:* it is also called as biometric based authentication. This includes the mechanism of fingerprint scan, iris scan etc.

© Springer International Publishing AG 2017
A. Basu et al. (Eds.): IHCI 2016, LNCS 10127, pp. 27–38, 2017.
DOI: 10.1007/978-3-319-52503-7_3

Most systems use knowledge based authentication mechanism as it balances the usability, deployability and security issues. But the "password problem" [14] associated with alphanumeric passwords are expected to comply with two basic conflicting requirements-one is associated with usability and the other is related to security aspects.

1. *Usability aspects:* passwords should be easy to remember.
2. *Security aspects:* passwords should be secure, should be hard to guess; they should be changed frequently, and should not be same for any two accounts of the same user; they should not be written down or stored in plain text.

Meeting both of these requirements is the main challenge here. For this reason, users tend to choose and handle alphanumeric passwords very insecurely. Research has shown that when users fail to recall a password, they are often able to recall some parts of it correctly. But partially correct password cannot authenticate a user, only exact recall of complete password is required.

Various cognitive and psychological studies [10,12] indicate that pictures are much easier to remember than texts. This is the main objective behind graphical passwords which use images or shapes as a replacement for text. In a conventional graphical password scheme, a user selects several images as his/her password. During login the user has to click on the password images from a larger set of distractor images. If the user identifies the password images, that were selected in the registration phase and clicks on them, he/she is successfully authenticated [15].

However, clicking on images on a large, vertical screen may allow an observer to capture the password images. To address the above issue we propose a graphical password system which is resistant to the above said attack and is also resistant to spyware attack.

2   Related Work

Based on the authentication style, graphical password system can be broadly classified into three categories:

1. *Graphical passwords based on recognition:* In recognition based graphical password techniques, a user is asked to identify some images that he has selected during registration i.e. passface [4].
2. *Graphical passwords based on cued recall:* In cued recall based graphical password techniques, a user is asked to reproduce something that he has selected during registration phase i.e. passpoint [14], story [5].
3. *Graphical passwords based on pure recall:* In pure recall based graphical password techniques, a user is asked to redraw something that he has created during registration phase i.e. DAS [9], PassShapes [13].

Here we focus on graphical passwords based on recognition and based on cued recall rather than the password system based on pure recall.

A general approach to design a graphical password system is a challenge-response process. In this challenge-response process a user creates his/her password by selecting several images from a set of images. During the authentication phase, a challenge is thrown to the user by displaying some decoy image with a password image. In response, user has to successfully respond by identifying and clicking on the password image shown on the screen.

The first graphical password scheme proposed by Blonder [3], in which the user is asked to click on the approximate predetermined locations of a pre-decided graphical image in a correct order. The image can help users to recall their passwords and it is a better one as compared to recall alphanumerical passwords. However, the predetermined regions can be easily identifiable.

"Passpoint" system proposed by Wiedenbeck et al. [14] extended Blonder's approach by addressing some of its limitations. It allows arbitrary images instead of the predefined one and users can click on any place on the image to create their passwords. Here, the possible password space is quite large. However, there are some apparent points on the image which are usually chosen by users as their passwords. It makes the work easy for an attacker.

Based on the idea of passpoints, Suo [11] proposed a shoulder-surfing resistant password scheme in which the image is blurred except for a small focused area. Users can enter Y(for yes) and N(for no) to indicate that their click-point is within the focused area. If the click-points are a few, attackers can easily guess [7].

Based on the fact that humans can recall faces easier than other images Brostoff et al. proposed "passface" [4]. Here, users need to recognize and click on the face images that are selected in the registration phase. This procedure is repeated for several rounds. If the user correctly responds to each round, then he/she is successfully authenticated.

Similar to passface, Dhamija et al. [6] proposed a graphical authentication scheme which uses abstract images and concrete photographs instead of faces. For authentication, user is required to click on the password images from a set of decoy images and password images.

By utilizing the properties of convex hull, Sobrado et al. [15] proposed a protocol in which the system will show a number of graphical icons. For authentication, user identifies the pass icons, and then mentally forms a convex hull of the pass icons and clicks on a random point inside the convex hull.

Based on two protocols DAS [9] and Story [5], Gao et al. [8] proposed a new user friendly shoulder-surfing resistant scheme, called CDS. Here in password creation process a user selects several images as passwords and remembers them with their order of selection. During login, the user has to draw a curve across the password images orderly without lifting the stylus. However, its small theoretical password space and hotspots issue make it vulnerable to brute force attack and dictionary attack [7].

Combining text with graphic, Zheng et al. [16] proposed a hybrid password authentication scheme which uses shapes of strokes on the grid as the origin passwords and allows the users to login through text passwords. The basic idea of this is to think some personal shape, map from this shape to text with strokes of the shape and a grid with text.

PassShapes [13] proposed by Weiss et al., authenticate users to a computing system by drawing simple geometric shapes constructed of an arbitrary combination of eight different strokes.

3   SG-PASS System

Our proposed scheme is a challenge-response mechanism and it uses approximately hundreds of graphical icons or arbitrary images shown in a window on the screen. In a challenge a user must identifies his or her password icons out of hundreds of displayed icons. Instead of direct input, the user responds to the challenge by entering keys from the keyboard keeping in mind that the shape formed by password icons on the screen matches with the shape of the entered keys of the keyboard. Two or three challenges are presented in sequence and if the user successfully responds to each challenge by entering the correct shape from the keyboard, then he or she is authenticated. It is based on recognition, an easier memory task than recall and users may create a story for sequence retrieval. In the following sections we describe the design and implementation in more detail.

3.1   Registration Phase

Our system uses a large set of images that are partitioned into different types for example, the images of sceneries, flowers, animals etc. A user can also add images of his/her choice. In registration phase, system will display a window consisting of approximately hundreds of images for creation of password. If a user feels uncomfortable with the images provided by the system, he/she can change that image window of his/her choice by adding different types of image. To create a password, the user chooses several images or icons from the window and also takes care about the order of selection of password images. Because, here the order of password images plays a vital role in authentication phase. The user has to remember the password images and their order that are selected by him/her. By creating a story a user may remember the order of password images.

3.2   Authentication Phase

In authentication phase, the password window containing same set of images but randomly permuted, is displayed to the user. These images include both password images and decoy images. User has to recognize the pre-selected password images and mentally draws a geometrical shape by connecting the password images in their respective order. Then the user enters keys anywhere from the keyboard [1] keeping in mind that the entered keys reflect the geometrical shape

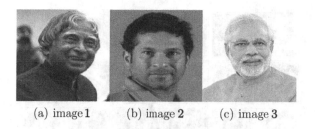

(a) image 1   (b) image 2   (c) image 3

Fig. 1. Password images in their order

formed by password images displayed on the screen. This process of challenge-response is repeated more than once and the exact number depends on the system administrator. When the user has responded to a challenge (either correctly or incorrectly), another challenge comes, and this process continues until all challenges have been completed. The images are arranged randomly inside the window each time it is displayed on the screen, so the password images move to new positions. A brief outline of the authentication steps for the proposed SG-PASS scheme is given below.

Authentication Steps

1. Identify all the password images from the set of images shown in the window screen. Now their relative positions are located.
2. Map the bottom most password image to the lowest row of the keyboard.
3. Find the row difference between top most password image and bottom most password image.
4. Keeping in mind, the total number of rows in password window to the total number of rows in keyboard (only four), map the topmost password image to the keyboard.
5. Then map the rest password image, accordingly.
6. After the process of mapping, enter the keys as per the order of preselected password images, such that the entered keys reflect an approximate shape formed by password images.

3.3 Discussion

Figure 2 shows a password window which has the password images of Fig. 1 and some decoy images (here we are considering only the faces of celebrities, user can also change the window so that one window may contain different types of images i.e. scenery, flowers, animals etc.). User can identify the password images and try to map the shape by entering the keyboard characters like "TCB" or "EZC" or "YVN" as "TCB", "EZC" are reflecting an approximate shape formed by the password images. But the system cannot accept "TVN" as password because it is not reflecting the shape formed by password images (Figs. 3, 4 and 5).

Fig. 2. Window1

Fig. 3. TCB, a valid response

In the prototype we have taken only three images as password images. One can take more than that also but the minimum number is three. During authentication only three images are considered at a time. Assume that, one user has selected four images as password image i.e. image1, image2, image3, image4. In authentication, the user will consider the first three images, that is, image1, image2, image3 and selects three keys from the keyboard accordingly. After that, in the same round, the user will take the next three images that is, image2, image3, image4 and selects three keys as per image2, image3, image4. Thus, the user will have to enter six keys at a time for password size of four.

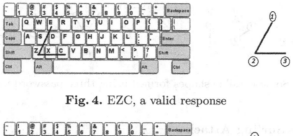

Fig. 4. EZC, a valid response

Fig. 5. TVN is not a valid response

4 Security and Usability Analysis

In this section, different security and usability factors of SG-PASS scheme is analyzed.

4.1 Mouse-Loggers

Mouse-loggers are used to record the click position and trajectory of the mouse. It can crack the password schemes which use mouse for input information [7]. Mouse-loggers are not a threat to our proposed scheme as we are giving information to the system using the keyboard not using mouse.

4.2 Keystroke-Loggers

The main idea of our proposed scheme is making a shape based password using the images and text input. The text characters that user will input are different in different login session. This mechanism can prevent key-logger attack. If an attacker records the input characters then he would get nothing about the password images.

4.3 Accidental Login

Assume that the number of password image is three and only two round of challenge response is there. Using three password images many different shapes are possible. If we consider only ten distinct shapes like in Fig. 6, then the probability of successful accidental login can be computed as below.

As we are considering the order, for each image six different orders are there (we are taking three password images in the prototype, it gives $3! = 6$ different types of permutation). So, using three password images there are 60 different passwords. For two round of challenge-response, the probability of successful accidental login becomes $(1/60)*(1/60) = 0.00028$ which is quite low.

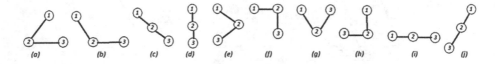

Fig. 6. Some possible shapes formed using three password images

4.4 Shoulder-Surfing Attack

The login process does not reflect the password images directly. Attacker can get the shape formed by password images by observing the text characters entered by the user. But, that shape cannot reveal the password images because for a single geometrical shape there are many combinations of password image. Consider the password window in Fig. 2, here many corresponding shapes can be formed by characters "TCB", some are given in Fig. 7.

Fig. 7. Corresponding shape for characters "TCB"

In addition as the password images changing their positions randomly inside the password window, a human observer cannot remember exact positions of all the images in a particular window.

4.5   Brute Force Search

The general mechanism to defend brute-force search is to increase the password space. Assuming that there are hundreds of images in a password window, the password space can be calculated as C(100,3) = 161700

However, it is difficult to carry out brute-force search in graphical passwords compared to alphanumeric passwords because computers spend considerable amount of time for identifying the password images in a password window.

4.6   Phishing

Phishing is a difficult one for graphical password scheme as compared to alphanumeric passwords. In the alphanumeric password scheme attackers need not require to know anything about user's password or theory behind authentication process. The fake website will record user name and password as entered by user. However, in graphical password system the attacker must know how the authentication process works and it is different for different graphical password scheme. In our scheme, the password window is different for different users as one can add images of his/her choice and users are not giving direct input for passwords. So, it is a difficult task for attackers to get passwords through phishing.

4.7   Usability Analysis

Fifteen participants including university students and non-technical staff took part in the evaluation process of SG-PASS scheme. After training, they took 27.6 s on average for each round of SG-PASS and found that it is easy to remember and a simple one. For the target of ten correct login, five participants accomplished in first ten attempts. However, the remaining participants reach the target with one to three extra attempts. The details are given in Table 1. The percentage of successful login attempts out of total login attempts by all users is found to 89 percent.

5   Comparison with Other Recognition Based Graphical Systems

In Sect. 2, we discussed various graphical password systems that use the recognition technique for authentication. We compare our system, SG-PASS with these password systems in terms of security and usability strength.

From Table 2, we can conclude that only two schemes CHC [15] and Zheng [16] can prevent both spy-ware and shoulder-surfing attacks. However, the distribution of response points are not uniform (more concentrated at center) in CHC, which makes the attacker's work easy. In Zheng et al.'s methodology [16], users have to remember exact locations of the grid cells like DAS, which puts an extra memory burden. In a nutshell, the comparative results are presented in Table 2. It shows SG-PASS has relatively higher security and usability features

Table 1. Details of usability analysis

Users	Avg. login time in sec	Unsuccessful attempt
User1	30	2
User2	25	3
User3	23	0
User4	22.4	0
User5	25	2
User6	26.1	1
User7	27	0
User8	27.3	1
User9	30	2
User10	28.2	0
User11	33.5	2
User12	29	0
User13	35	2
User14	25.3	1
User15	27	2

Table 2. Comparison with other methodology

Schemes	Spy-ware resistant	Shoulder-surfing resistant	Comment
Blonder [3]	N	N	Number of predefined regions are small and easily identifiable
Passpoint [14]	N	N	Care must be taken to eliminate hot spots
Suo [11]	N	N	Attackers can easily guess if few click-points are used
Passface [4]	N	N	Face images are clearly visible
Dhamija [6]	N	N	A recondite picture is hard to remember
CHC [15]	Y	Y	Distribution of the response points are not uniform [2]
CDS [8]	N	Y	Password space is very less
Zheng [16]	Y	Y	Users have to remember exact locations of the grid cells like DAS [9]
SG-PASS	Y	Y	Simple, easy to remember and resist all the attacks discussed in Sect. 4

compared to existing graphical password schemes. However, like other schemes it also suffers from intersection issue. An intersection attack is possible when the attacker is able to record the password window and the keys for multiple sessions.

6   Conclusion and Future Work

The contribution of this paper is the design of a graphical password scheme that extends the challenge-response model to resist spy-ware and shoulder-surfing attacks. Users can create a valid graphical password easily and quickly but face some difficulty in learning their passwords. This scheme is easy to execute and more secure and usable as compared to other graphical password approaches. It provides a simple and intuitive technique for users to authenticate. However, like other graphical password system, the issues in this system is the intersection analysis. To overcome this issue, we plan to design a more advanced system without compromising the security and usability aspects.

References

1. Ameer, D., Al-Absi, A.A., Mohammed, A.O., Habbal, A.M.M., Hassan, S.: Anywhere on-keyboard password technique. In: 2010 IEEE Student Conference on Research and Development (SCOReD), pp. 159–163. IEEE (2010)
2. Asghar, H.J., Li, S., Pieprzyk, J., Wang, H.: Cryptanalysis of the convex hull click human identification protocol. Int. J. Inf. Secur. **12**(2), 83–96 (2013)
3. Blonder, G.E.: Graphical password, uS Patent 5,559,961, 24., September 1996
4. Brostoff, S., Sasse, M.A.: Are passfaces more usable than passwords? a field trial investigation. In: People and Computers XIV Usability or Else!, pp. 405–424. Springer (2000)
5. Davis, D., Monrose, F., Reiter, M.K.: On user choice in graphical password schemes. In: USENIX Security Symposium, vol. 13, p. 11 (2004)
6. Dhamija, R., Perrig, A.: Deja vu-a user study: using images for authentication. In: USENIX Security Symposium, vol. 9, p. 4 (2000)
7. Gao, H., Jia, W., Ye, F., Ma, L.: A survey on the use of graphical passwords in security. J. Softw. **8**(7), 1678–1698 (2013)
8. Gao, H., Ren, Z., Chang, X., Liu, X., Aickelin, U.: A new graphical password scheme resistant to shoulder-surfing. In: 2010 International Conference on Cyberworlds (CW), pp. 194–199. IEEE (2010)
9. Jermyn, I., Mayer, A.J., Monrose, F., Reiter, M.K., Rubin, A.D., et al.: The design and analysis of graphical passwords. In: Usenix Security (1999)
10. Shepard, R.N.: Recognition memory for words, sentences, and pictures. J. Verbal Learn. Verbal Behav. **6**(1), 156–163 (1967)
11. Suo, X.: A design and analysis of graphical password (2006)
12. Suo, X., Zhu, Y., Owen, G.S.: Graphical passwords: a survey. In: 21st Annual Computer Security Applications Conference (ACSAC'05), pp. 463–472. IEEE (2005)
13. Weiss, R., De Luca, A.: Passshapes: utilizing stroke based authentication to increase password memorability. In: Proceedings of the 5th Nordic Conference on Human-Computer Interaction: Building Bridges, pp. 383–392. ACM (2008)

14. Wiedenbeck, S., Waters, J., Birget, J.C., Brodskiy, A., Memon, N.: Passpoints: design and longitudinal evaluation of a graphical password system. Int. J. Hum Comput Stud. **63**(1), 102–127 (2005)
15. Wiedenbeck, S., Waters, J., Sobrado, L., Birget, J.C.: Design and evaluation of a shoulder-surfing resistant graphical password scheme. In: Proceedings of the working conference on Advanced visual interfaces, pp. 177–184. ACM (2006)
16. Zheng, Z., Liu, X., Yin, L., Liu, Z.: A hybrid password authentication scheme based on shape and text. J. Comput. **5**(5), 765–772 (2010)

O-PrO: An Ontology for Object Affordance Reasoning

Rupam Bhattacharyya[✉], Zubin Bhuyan,
and Shyamanta M. Hazarika

Biomimetic and Cognitive Robotics Lab, Department of Computer Science
and Engineering, Tezpur University, Tezpur 784028, India
{rupam15, zubin, smh}@tezu.ernet.in

Abstract. Object affordances provide useful information related to understanding of human activities. The aim of this paper is to create an ontology for object affordance reasoning that can be shared across different assistive robots operating within the household domain. A novel ontology called O-PrO (Object Property Ontology) consisting of 61 household objects is presented. The ontology can be used for computing cognitive and semantic object affordances.

Keywords: Object affordance · Ontology · Assistive robot

1 Introduction

A key factor for successful operation of an assistive robot in household environment is its ability to process contextual information. A variety of contextual information from different sources is available. Difficulty lies in picking up the appropriate contextual information in a limited amount of time and using it to build the robotic controller. Our primary interest lie in representation and efficient processing of such contextual information related to object affordances involving different household activities. Utilization of object affordance helps one in the following ways: a. reduction in maintaining the complex internal representation [4] of agent-environment interactions and b. understanding the behavior of humans.

1.1 Object Affordance: Definition and Its Varieties

Object affordance is defined as "properties of an object that determine what actions a human can perform on them"; as defined in the seminal paper by Gibson [2]. Figure 1 illustrates the idea of object affordance vis-à-vis human intention and activities involving objects. Our focus is on the shaded box i.e. *object properties*. Table 1 shows the types of affordances conceived by researchers in the recent years. For a more elaborate discussion on affordances readers are referred to [6, 13, 14].

O-PrO targets only the semantic and cognitive object affordances. Other affordances such as spatial, temporal and social affordance do not focus on the traditional definition of object affordances. Hence, we concentrate on the object properties rather than the temporal or spatial aspects.

© Springer International Publishing AG 2017
A. Basu et al. (Eds.): IHCI 2016, LNCS 10127, pp. 39–50, 2017.
DOI: 10.1007/978-3-319-52503-7_4

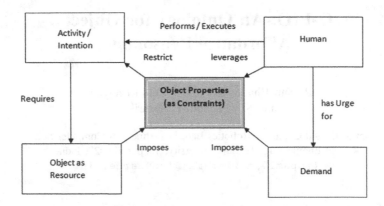

Fig. 1. Abstract model for understanding object affordance

Table 1. Types of affordances introduced in recent literature

Affordance type	Reference	Research field	Evaluation domain
Semantic affordance	Koppula et al., 2013 [7]	Computer vision, robotics	Video Data (CAD 120)
Temporal affordance	Koppula et al., 2013 [6]	Computer vision, robotics	Video Data (CAD 120)
Spatial affordance	Koppula et al., 2013 [6]	Computer vision, robotics	Video Data (CAD 120)
Cognitive affordance	Sarathy and Scheutz, 2016 [13]	Knowledge representation	N.A.
Social affordance	Shu et al., 2015 [14]	Computer vision	Video data

2   Motivation

Most of the existing literature on object affordance from video data does not consider the deeper visual reasoning of the human mind. Contextual information representation using ontology has paved the way for building such robust knowledge based systems. Two obvious questions which arise with regard to such systems are:

1. How to construct and represent the entities of the domain in an ontology, and
2. How to use the ontology for improving reasoning tasks through object affordance.

Three motivating scenarios are discussed below to highlight the research gap:

Scenario 1: This example is taken from [13] to underscore that their reasoning is insufficient related to object affordance. Their system is unclear on how it addresses when Julia says to the robot: "Bring me something clean I can use to cut this tomato."

Our analysis: The part affordances of the object ("something") have to be "sharpEdge" and "graspable" at the same time. [13] use physical features in their reasoning process. However, we are not able to find out any relevant source of physical features.

Such features might help the robot to identify that the object weight and volume has to be small enough to lift that object properly. Human can perform deeper visual reasoning after observing an object. Only visual attributes (for object recognition) by the computer vision module cannot help one to arrive at such kind of reasoning.

Scenario 2: Let us consider two high level activities "make tea" and "make cereal" which are common in our daily indoor activities in the kitchen environment.

Our analysis: Table 2 shows different semantic affordance ids involved in these two activities. Although the objects involved in the two sub-activities are different, but object affordances of these objects are similar. We argue that objects belonging to the same category possess similar object affordances. For example, graspable affordance has a definite object set. Objects from such a set can be interchanged increasing the adaptability of robots in recognizing different human intentions or activities [9].

Table 2. Object affordances involved in these activities

Top level (high-level activity)	Mid level (sub-activity)	Low level actions	Object affordances present in the scene
Make tea	Pouring hot water	Grasp kettle, pour water into cup, release kettle	Graspable, pour-to, pourable, placeable
Make cereal	Pouring milk	Grasp container, pour milk into bowl, place container	Graspable, pour-to, pourable, placeable

Scenario 3: Suppose, a person in the kitchen environment is performing a high-level activity called "microwaving food".

Our analysis: In [7], "openable" and "closeable" affordance ids are annotated with the object "microwave oven" in the high-level activities like microwaving food and cleaning objects. The part affordance of the door or the geometric mapping of the microwave oven activates its affordances like "openable" and "closeable". Metric spatial relations like nearby, on top of, between objects or human-objects cannot single-handedly identify such specific object affordance ids. We believe that they are not utilizing the full potential of object affordances.

3 Related Work

3.1 Existing Ontology for Object Affordances

Research conducted by Varadarajan and Vincze [16] can be considered as a significant contribution in the area of object affordances. The affordance knowledge ontologies provided through the affordance network (AffNet) for household articles inspires our work. They highlighted problems with the state of the art systems like KnowRob [15] or ConceptNet [3]. We have AffNet 2.0 as a source of semantic information related to affordances.

Object affordance evolved from psychology research. [11, 12] talked about ontologies for affordances; but there was no proper ontology for evaluating their work. This motivates us to create an open source ontology to be used by real time robotic controllers for human indoor activity/intention understanding.

3.2 Knowledge Based Approaches Towards Semantic and Cognitive Affordances

Ontologies play a fundamental role in providing formal models of domain knowledge which can be used by intelligent agents like assistive robots [5]. A knowledge graph consisting of visual, physical and categorical attributes to understand the deeper human visual reasoning have been used [17]. This image based evaluation framework considers only 40 objects (excludes natural objects) for the knowledge graph construction. There is no algorithm mentioned about how the online sources (e-bay and amazon) are queried for physical attributes. One attribute is shared by multiple category of objects which increase the inference time to find out the appropriate object affordance label. There is no clear explanation about how structural attribute or part affordance influences the overall affordance of an object. Within cognitive object affordance, no ontology has been used [13]; rather predicate-style representation is done for context, percepts, and affordances.

4   O-PrO: Ontology for Object Properties

We have used Protégé 5.0 to author our ontology, O-PrO. In this section, we discuss the physical objects which have been described and modeled in O-PrO. This ontology is freely available[1] on GitHub under the MIT license.

4.1   Objects

In a generic household domain, we would like to categorize the objects in two classes, viz., (i) natural objects and (ii) artifacts. We have considered total 61 objects involving both natural objects and artifacts. 51 objects are chosen from the standard RGBD object dataset [8]. Rests of the objects are chosen from [17], which are most commonly used in a variety of indoor human activities.

4.2   Object Properties

We have considered three types of object attributes namely, physical, categorical and structural attributes. Physical attributes comprise of volume and weight of the objects. In order to extract meaningful properties related to the functionality of the object, we

---

[1] O-PrO can be downloaded from the GitHub repository, https://git.io/viO6Q.

have to consider its structural attributes such as its constituent parts and its geometric shape. The physical and structural attributes have been considered in qualitative form for concise representation. Three enumerated OWL classes viz., QualitativeWeight, QualitativeVolume, QualitativeConcavity have been defined in the ontology. The product specifications from the previously mentioned sources have been used to assign qualitative values to the objects. The QualitativeMeasure class and its sub-classes have been described in Figs. 2 and 3. Additionally, a Shape class has been defined to assign approximate geometric shape to the objects or its parts. Basic shapes such as cuboid, sphere, pole, disc, cylinder and wedge form this enumerable class. Figures 4 and 5 illustrate the Shape class. Tables 3 and 4 shows the attributes used to describe the 61 objects in the household domain.

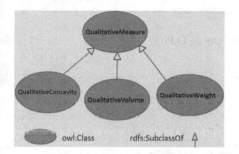

QualitativeWeight	≡ {LightWeight, MediumWeight, HeavyWeight}
QualitativeVolume	≡ {LowVolume, MediumVolume, HighVolume}
QulatitiveConcavity	≡ {Nil, LowConcavity, MediumConcavity, HighConcavity, LowConcavityInv, MediumConcavityInv, HighConcavityInv }

Fig. 2. QualitativeMeasure class and its subclasses

Fig. 3. DL axiom for subclasses of QualitativeMeasure.

Table 3. Attribute categories used for object description

Attribute category	Attribute name
Physical	Weight, Volume
Structural	Sub-component, hasConcavity
Categorical	These attributes are arranged in an Is-A hierarchy (See Fig. 6 for details)

Table 4. Object-properties for describing household objects

Object-property	Range	Description
hasConcavity	QualitativeConcavity	Concavity or convexity of an object in context of its most common usage
hasVolume	QualitativeVolume	Qualitative volume of the object from the above mentioned sources
hasWeight	QualitativeWeight	Qualitative weight of the object from the above mentioned sources
hasGeometricShape	Shape	Approximate shape of the object or its parts
hasPart	Component	Constituent parts of the object. The sub-properties of this property can be visualized from the Fig. 6

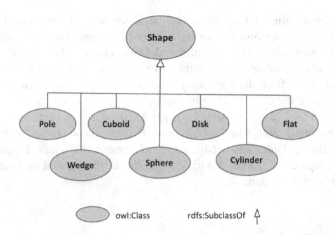

Fig. 4. Shape class hierarchy in O-PrO.

```
<EquivalentClasses>
        <Class IRI="#Shape"/>
                <ObjectOneOf>
                        <NamedIndividual IRI="#Cuboid"/>
                        <NamedIndividual IRI="#Flat"/>
                        <NamedIndividual IRI="#Pole"/>
                        <NamedIndividual IRI="#Disk"/>
                        <NamedIndividual IRI="#Cylinder"/>
                        <NamedIndividual IRI="#Wedge"/>
                        <NamedIndividual IRI="#Sphere"/>
                </ObjectOneOf>
</EquivalentClasses>
```

Fig. 5. OWL definition of the nominal Shape class from O-PrO.

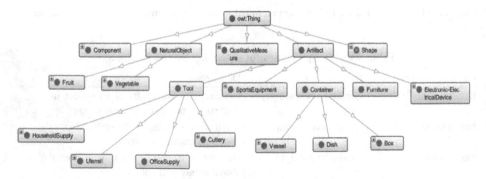

Fig. 6. O-PrO visualized in Protégé OntoGraf.

5 Domain Knowledge Infrastructure

5.1 Available Semantic Resources

Resources for physical attributes: Physical attributes considered in our ontology are volume and weight. Following resources were helpful in extracting this attribute.

Amazon[2]. This online web resource contains varieties of object items as their products available for their customers. The textual product specification is valuable information for perceiving the size and weight of household objects.

Flipkart[3] and eBay[4]. These are two online resources similar to amazon. However, the issue with these resources is that they do not contain product specifications about natural objects. Apart from this, several items do not have specific product specifications. There are numerous online shopping sites available. We are restricting ourselves to these three sites for constructing our ontology, O-PrO.

Resources for categorical attributes: The following web resource assisted us in categorizing the 61 objects we have considered.

ImageNet[5]. A widely known resource for image data processing is ImageNet [1]. It is a database consisting of images engineered in agreement with the WordNet [10] hierarchy. Anyone can explore the online content of ImageNet for identifying categories of object items.

Resources for structural attributes: To identify structural attributes, one useful online web resource is the *AffNet 2.0* website[6]. Part affordances and geometric mapping are also mentioned for a variety of objects. We have not considered the material attributes of the objects given in AffNet 2.0.

General resource for overall object properties: Various web resources give us useful information about overall object properties.

ConceptNet[7]. ConceptNet is a freely usable commonsense knowledge base. From [3], ConceptNet is "a semantic network containing lots of things computers should know about the world, especially when understanding text written by people." The problem of this multilingual knowledge base is that it contains additional information which are not relevant for object affordance reasoning process. The advantage of this resource is that all the three object attributes information can be found in ConceptNet. ConceptNet have a number of vulnerabilities as a knowledge source [16].

---

[2] http://www.amazon.com/.

[3] https://www.flipkart.com/.

[4] http://www.ebay.in/.

[5] http://image-net.org/.

[6] http://theaffordances.appspot.com/.

[7] http://conceptnet5.media.mit.edu/.

WordNet[8]. The WordNet [10] relationships cover only linguistic dependencies. We have queried the WordNet with the word *microwave oven*. But, there is no relationship in WordNet linking the door of microwave oven with the overall structure of it (microwave oven). Nevertheless, it provides useful information about shape/size of various object items which cannot be found in other sources.

Wikipedia[9]. It helps to find out size and weight of natural objects which are missing from previous resources. Due to its immense familiarity among readers; we are not going to talk about this knowledge resource further.

5.2 Knowledge Acquisition Phase

We have manually collected all the relevant information from the above mentioned resources to construct O-PrO. The procedure followed during this phase is given below:

```
procedure ExtractObjectProperties
  STEP 1: Initialize two variables objectLimit=61, coun-
          ter=1;
  STEP 2: if(counter ≤ 51) goto STEP 3 else goto STEP 14
  STEP 3: Extract object name from RGBD dataset [8].
  STEP 4: Initialize variable objectName with extracted
          name.
  STEP 5: if (objectName ∉Natural object) goto STEP 6
          else set variable FLAG=1 and goto STEP 10
  STEP 6: Extract data range for variable volume and
          weight for objectName.
  STEP 7: Initialize variable count=1, sumVolume=0 and
          sumWeight=0.
  STEP 8:
        while (count ≤ 3) do
          switch (count)
            Case 1: Query the amazon website with object-
                    Name.
                    Initialize variable volumeObjectName and
                    weightObjectName.
                    sumVolume=sumVolume+sizeObjectName
                    sumWeight=sumWeight+weightObjectName
            Case 2: Query the flipkart website with ob-
                    jectName
                    Initialize variable volumeObjectName and
                    weightObjectName.
                    sumVolume=sumVolume+sizeObjectName
                    sumWeight=sumWeight+weightObjectName
```

---

[8] http://wordnetweb.princeton.edu/perl/webwn.

[9] https://www.wikipedia.org/.

```
                Case 3: Query the eBay website with objectName
                        Initialize variable volumeObjectName and
                        weightObjectName.
                        sumVolume=sumVolume+sizeObjectName
                        sumWeight=sumWeight+weightObjectName
                End of switch statement
                count++
            End while
            Initialize volumeObjectName=sumVolume/3
            Initialize weightObjectName=sumWeight/3
    Step 9: Extract structural attribute from AffNet 2.0
            Initialize hasGeometricShape variable and partOf
            relation.
            Verify the values with ConceptNet 5.0
    Step 10: if(FLAG==1)
                    Query the Wikipedia website for the volume
                    and weight of the natural object.
                    Initialize variable volumeObjectName and
                    weightObjectName.
                else do nothing
    STEP 11: if (FLAG==1)
                    Query the ConceptNet and WordNet websites
                    for the structural attribute of the natural
                    object.
                    Initialize hasGeometricShape variable and
                    partOf relation.
                else do nothing
    STEP 12: Extract categorical attribute from Imagenet.
                Initialize hasCategory_objectName variable
                with the result of the previous step.
    STEP 13: counter++
    STEP 14: FLAG=0
    STEP 15: if (counter > 51 and counter ≤ 61)
                    Extract object name from [17]
                    Goto STEP 4
                else if (counter > 61) goto STEP 16
                else goto STEP 3
    STEP 16: Stop
End procedure
```

5.3   Envisaged Application-vis-à-vis Scenario

Figure 7 illustrates a typical object described in O-PrO. We consider the example of a microwave oven and its properties. Any query to O-PrO, will retrieve such an object with its properties. Vision based grounding of object properties have not been done.

Three scenarios described in Sect. 2 highlights the research gap in object affordance reasoning. Our solution can be considered as a unified approach which is able to

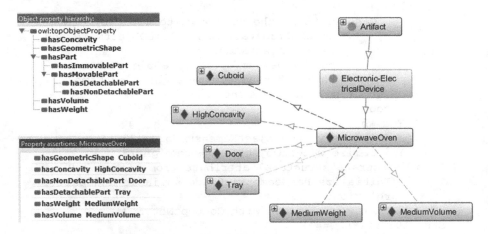

Fig. 7. An artifact instance visualized in Protégé OntoGraf

consider all three scenarios at the same time. The information related to different parts of the object (MicrowaveOven in Fig. 7) and its shape information is readily available for the reasoner. Similarly, physical features (MediumWeight and MediumVolume) of the microwave are also available. From Fig. 7; we observe that an artifact with high concavity and cuboid shape makes the microwave *containable*. All the artifacts having similar structural attributes indicate the possibility of an object set for a specific affordance. O-PrO also provides information related to detachable and non-detachable part of the objects. Thus, a vast range of action possibilities (like *openable* and *closeable* in Scenario 3) can be inferred. An assistive robot using our ontology for the activity "microwaving food" is able to act more intelligently by accessing different actions possibilities through various object properties of the microwave.

Readers can observe the difference between O-PrO with other ontologies from the scenarios mentioned in Sect. 2: a. O-PrO can provide three object properties namely physical, categorical and structural. AffNet 2.0 [16] only provides part affordances of object items by considering structural and material attributes. b. We consider both natural object and artifacts in our ontology. These objects (total 61 object items) are chosen based on their relevance in human daily indoor activities. AffNet 2.0 [16] does not consider natural objects and some key household appliances.

6   Conclusion

O-PrO (Object Property Ontology) consisting of 61 household objects is presented. Three types of object attributes were considered while creating the ontology. The ontology can be shared by different assistive robots operating in household environments to quickly gather relevant information for object affordance reasoning. The three scenarios put forth in Sect. 2 illustrates the need for the three object properties viz., a. physical (size and weight), b. categorical and c. structural. Physical and categorical attributes are used in [17]; however, use of categorical attributes and their benefits is not clear. Further, their image based evaluation allows them to have ample time to extract different web resources for object affordance reasoning. In the AffNet 2.0 ontology [16], part affordances and geometric mapping for natural objects and various household appliances like microwave oven, vacuum cleaner, pliers, etc. are not considered. The proposed ontology can be queried to extract relevant information in real time. Object affordance labels are not kept as the object's property since object affordance may change frequently in a video based framework.

Acknowledgement. This research work is supported by UGC, Government of India under SAP Level-II. Financial assistance received under DST-UKIERI DST/INT/UK/P-91/2014 is gratefully acknowledged.

References

1. Deng, J., Dong, W., Socher, R., Li, L.J., Li, K., Fei-Fei, L.: Imagenet: a large-scale hierarchical image database. In: IEEE Conference on Computer Vision and Pattern Recognition, CVPR 2009, pp. 248–255. IEEE (2009)
2. Gibson, J.J.: The ecological approach to the visual perception of pictures. Leonardo **11**(3), 227–235 (1978)
3. Havasi, C., Speer, R., Alonso, J.: Conceptnet 3: a flexible, multilingual semantic network for common sense knowledge. In: Recent Advances in Natural Language Processing, pp. 27–29. Citeseer (2007)
4. Horton, T.E., Chakraborty, A., Amant, R.S.: Affordances for robots: a brief survey. Avant **3**(2), 70–84 (2012)
5. Knublauch, H., Fergerson, R.W., Noy, N.F., Musen, M.A.: The Protégé OWL plugin: an open development environment for semantic web applications. In: McIlraith, Sheila, A., Plexousakis, D., Harmelen, F. (eds.) ISWC 2004. LNCS, vol. 3298, pp. 229–243. Springer, Heidelberg (2004). doi:10.1007/978-3-540-30475-3_17
6. Koppula, H.S., Saxena, A.: Physically grounded spatio-temporal object affordances. In: Fleet, D., Pajdla, T., Schiele, B., Tuytelaars, T. (eds.) ECCV 2014. LNCS, vol. 8691, pp. 831–847. Springer, Heidelberg (2014). doi:10.1007/978-3-319-10578-9_54
7. Koppula, H.S., Gupta, R., Saxena, A.: Learning human activities and object affordances from RGB-D videos. Int. J. Robot. Res. **32**(8), 951–970 (2013)
8. Lai, K., Bo, L., Ren, X., Fox, D.: A large-scale hierarchical multi-view RGB-D object dataset. In: 2011 IEEE International Conference on Robotics and Automation (ICRA), pp. 1817–1824. IEEE (2011)
9. Liu, R., Zhang, X.: Understanding human behaviors with an object functional role perspective for robotics. IEEE Trans. Cogn. Develop. Syst. **8**(2), 115–127 (2015). doi:10.1109/TAMD.2015.2504919
10. Miller, G., Fellbaum, C.: Wordnet: an electronic lexical database (1998)
11. Moralez, L.A.: Affordance ontology: towards a unified description of affordances as events. Res. Cogitans **7**(1), 35–45 (2016)
12. Sanders, J.T.: An ontology of affordances. Ecol. Psychol. **9**(1), 97–112 (1997)
13. Sarathy, V., Scheutz, M.: Cognitive affordance representations in uncertain logic. In: Fifteenth International Conference on the Principles of Knowledge Representation and Reasoning (2016)

14. Shu, T., Ryoo, M., Zhu, S.C.: Learning social affordance for human-robot interaction. In: International Joint Conference on Artificial Intelligence (IJCAI) (2016)
15. Tenorth, M., Kunze, L., Jain, D., Beetz, M.: Knowrob-map-knowledge-linked semantic object maps. In: 2010 10th IEEE-RAS International Conference on Humanoid Robots, pp. 430–435. IEEE (2010)
16. Varadarajan, K.M., Vincze, M.: Afrob: the affordance network ontology for robots. In: 2012 IEEE/RSJ International Conference on Intelligent Robots and Systems, pp. 1343–1350. IEEE (2012)
17. Zhu, Y., Fathi, A., Fei-Fei, L.: Reasoning about object affordances in a knowledge base representation. In: Fleet, D., Pajdla, T., Schiele, B., Tuytelaars, T. (eds.) ECCV 2014. LNCS, vol. 8690, pp. 408–424. Springer, Heidelberg (2014). doi:10.1007/978-3-319-10605-2_27

Emotion Recognition from Facial Expressions of 4D Videos Using Curves and Surface Normals

Sai Prathusha S.[1](✉), Suja P.[1], Shikha Tripathi[2], and Louis R.[3]

[1] Department of Computer Science and Engineering,
Amrita School of Engineering, Amrita Vishwa Vidyapeetham,
Amrita University, Bengaluru, India
prathyureddy491@gmail.com, p_suja@blr.amrita.edu
[2] Department of Electronics and Communication Engineering,
Amrita School of Engineering, Amrita Vishwa Vidyapeetham,
Amrita University, Bengaluru, India
t_shikha@blr.amrita.edu
[3] CPE Lyon Engineering School, Lyon, France
louisrapet@gmail.com

Abstract. In this paper, we propose and compare three methods for recognizing emotions from facial expressions using 4D videos. In the first two methods, the 3D faces are re-sampled by using curves to extract the feature information. Two different methods are presented to resample the faces in an intelligent way using parallel curves and radial curves. The movement of the face is measured through these curves using two frames: neutral and peak frame. The deformation matrix is formed by computing the distance point to point on the corresponding curves of the neutral frame and peak frame. This matrix is used to create the feature vector that will be used for classification using Support Vector Machine (SVM). The third method proposed is to extract the feature information from the face by using surface normals. At every point on the frame, surface normals are extracted. The deformation matrix is formed by computing the Euclidean distances between the corresponding normals at a point on neutral and peak frames. This matrix is used to create the feature vector that will be used for classification of emotions using SVM. The proposed methods are analyzed and they showed improvement over existing literature.

Keywords: Emotion recognition · Feature vector · Deformation matrix · Parallel curves · Radial curves · Surface normals · Classifier

1 Introduction

Emotions are important for humans to communicate and express their feelings to others and play a vital role in people's everyday life. Emotions in humans can be recognized by using textual information, voice, gestures, postures, facial expressions.

Emotion recognition can be divided into four different main approaches: 2D static using only one frame, 2D dynamic using a sequence a frames, 3D static using a unique

© Springer International Publishing AG 2017
A. Basu et al. (Eds.): IHCI 2016, LNCS 10127, pp. 51–64, 2017.
DOI: 10.1007/978-3-319-52503-7_5

mesh of 3D points, 3D dynamic uses a sequence of 3D meshes. Facial expressions are dynamical; hence observing the deformations in the sequences of 3D faces can help to improve accuracy in emotion recognition. Also 3D data is resistant to pose variations and lighting conditions. Hence, in this work 3D data is used in performing facial shape analysis for face recognition and expression recognition. Now-a-days, consumer 3D cameras provide low resolution sequences of 3D faces. Due to this improvement in 3D imaging, 4D datasets such as BU-4DFE, Hi4D-ADSIP, dynamic 3-D FACS dataset etc., are available.

In this paper, we present the work on 3D dynamic meshes. Two different approaches are proposed in this paper to resample the 3D faces by using profile curves and radial curves. The movements in faces are measured by using these curves and for that we used two frames, neutral frame and peak frame. The deformation matrix is formed by computing the distance between point to point on each of the corresponding curves of neutral frame and peak frame. Then, feature vector is formed by using deformation matrix which is used for classification. In the third method, surface normals are extracted on each and every point of the ten frames selected out of the total frames of a video sequence of BU-4DFE database. The deformation matrix is formed by calculating the Euclidean distances between the corresponding normals of a neutral frame and a peak frame. The feature vector which is used for classification is formed by using the deformation matrix. The Multi-Class Support Vector Machine (SVM) is used for classification.

2 Literature Survey

In this section, curve based representation of 3D faces and some existing approaches for expression recognition using 4D are discussed.

In [1], a new method to recognize facial expressions from 4D video sequences is proposed. The faces are represented by using radial curves emanating from their tips of noses. To calculate deformation across the sequences, a method called Deformation Vector Field based on Riemannian facial shape analysis which captures densely dynamic information from the entire face is used. The feature vector is formed by using this temporal vector field. For classification, Multi-class Random Forest Algorithm is used.

In [2], two different 3D face registration algorithms are discussed. In the first algorithm, TPS warping algorithm, the faces are non-linearly folded to establish registration. The second algorithm, ICP algorithm, is used in the case of rigid deformations. Different 3D face recognition techniques which are based on 3D shape features such as 3D point coordinates, profile curves, surface normals, 2D depth images, etc., are also discussed. Though surface normals are not used as features in the present techniques which are used in face recognition, they gave the best accuracy in face recognition. The techniques for the fusion of 3D shape features at the decision level such as voting schemes, rank-based combination rules, product rules, fixed rules, serial-fusion schemes are also discussed.

In [3], an automatic method for recognition of facial expressions from 4D videos is explained. Radial curves are used to represent faces and Deformation Scalar Field which gives the direction of geodesic paths constructed between pairs of respective radial curves of two faces which is based on Riemannian geometry is used to estimate deformation between two faces. As the scalar fields have high dimensionality, LDA is applied. Classification is done in two ways: (1) Dynamic HMM is learned on the features. (2) Mean deformation is calculated under a window and Multi-Class Random Forest Algorithm is applied.

In [4], another approach for recognizing facial expressions is explained. The face is represented by the collection of level curves. Level curves divide the face into normalized segments. The chamfer distances between normalized segments are used as spatio-temporal features for facial expression recognition. Universal back-ground modeling is used as classification algorithm.

The information of 3D facial shape plays a vital role in the systems that recognize 3D faces. In [5], the techniques used to represent the shapes of 3D faces are based on point coordinates, surface normals, statistical analysis of depth images, facial profiles. The classifiers for face recognition have been developed based on these techniques. A comparison between these techniques is made by using 3D RMA dataset. Results showed that point-cloud representation and LDA based analysis of depth images performs best. The development of multiple-classifier schemes to mix individual face classifiers which are based on the information of 3D facial shape in both parallel and hierarchical ways is also discussed.

In [6, 7], approach for recognizing emotions from 4D video sequences is proposed. In each frame of a BU4DFE database there are 83 annotated landmark points. Out of that 83 points, key points are taken and euclidean distance between those points are calculated both in the neutral frame and peak frame of a video sequence. The differences between the euclidean distances between the neutral frame and peak frame are taken as features and feature vector is formed and given to SVM and Neutral Network classifiers and tried with different kernels.

The majority of work done in automatic recognition of facial expressions is on 2D images. But, due to the inherent problems in pose and illumination variations while using 2D images, the use of 3D images and 4D videos in the research of automatic recognition of facial expressions has been increased. In this paper [8], developments in acquisition of 3D facial data and tracking, available databases of 3D/4D face images used for the analysis of facial expressions of 3D/4D face images, presently available systems used to recognize facial expressions that use 3D images or 4D videos, the problems encountered while recognizing expressions of 3D/4D face images which are considered for future research are discussed.

Facial expressions play a vital role in the interaction. So, the efficient recognition of facial expressions is considered as a challenge in automatic facial image analysis. But, most of the available databases have 2D static images and videos of posed facial behavior. But, as the acted and un-acted facial expressions differ in intensity, complexity and timing, an un-posed video with annotations is needed. In this paper [9], a database which consists of videos of 3D face images expressing facial expressions which are spontaneous in nature is presented. In this paper, the 3D spatio-temporal features during the facial expression, understanding how the dynamic motion of facial action units changes with pose, understanding the natural occurrence of facial action are discussed.

In this survey [10], the focus is made on the topics such as recognition of discrete expressions using 3D face images. The research made on the facial expression recognition using 3D facial data is summarized and the drawbacks of the present methods have been discussed. The difficulties towards implementing more reliable and mechanized methods for the recognition of facial expressions using 3D facial data have been discussed.

By summarizing the literature, the 3D dynamic facial emotion recognition has been used for accurate recognition of emotions. Various Feature extraction methods based on curves, surface normals, 3D point coordinates, 2D depth images are there. Most common used classifiers are Hidden Markov Model (HMM) and Support Vector Machine(SVM). The challenges in recognizing emotions are pose and frequent head movements, occlusion, subtle facial deformation, ambiguity and uncertainty in face motion measurement.

3   Proposed Method for Recognition of Facial Expressions

The block diagram of feature extraction methods is shown in Fig. 1. The pre-processing steps which are followed are mentioned. The three different methods followed for the feature extraction are mentioned. The classifier used is SVM.

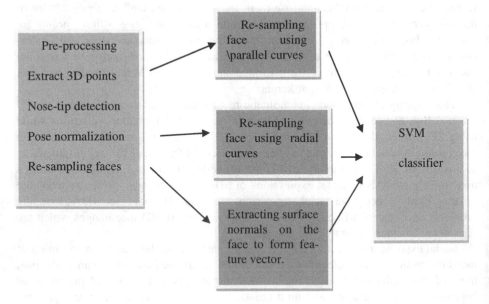

Fig. 1. Block diagram of feature extraction methods

3.1 Pre-processing

a. Nose detection

To perform the nose tip detection, most of the literature uses the highest point in 'z' as the nose tip. In the BU-4DFE database this method cannot be applied be-cause in few subjects, hair or mouth will be the highest points. The proposed procedure is as follows: (1) first, outliers are removed using mean and standard deviation. (2) To detect the nose tip a specific region is cropped with a sphere centered in the mean point. (3) After detecting the nose tip correctly, the Inertia matrix is calculated to align the face correctly. (4) Finally it is cropped using a cylinder.

This algorithm is used for the detection of the nose tip in the first frame. For the subsequent frames, the same region is used for detecting nose tips. The nose tip is detected correctly using this approach for the 60 subjects under consideration.

b. Face alignment using ICP

Based on the work in 3D emotion recognition, Iterative Closest Point (ICP) algo-rithm has been used to align all the faces and be able to compare them. The output of ICP algorithm will be a rotation matrix and a translation vector that minimize distance between the two points cloud. We decided to create a template by taking mean of 40 neutral faces. For each subject, emotion, frame, ICP is computed with this template.

To be sure that all the faces are correctly aligned, all the faces are translated to make the nose tip, the point $(0,0,0)$ simply by subtracting nose tip coordinate from all the points.

c. Re-sampling and final crop

As all faces are well aligned, the final crop and the re-sampling of the faces is done by fitting all the faces to a uniform grid and finally cropping all the faces with a cylinder which assures us that all frames of different profiles and emotions will have the same number of points and the point $(0, 0, 0)$ will always be the nose tip.

d. Peak detection

The database BU-4DFE provides 3D videos. Use of videos instead of only one image can provide more information by using the face motion to interpret emotions and perform a dynamic approach. The peak frame is detected out of the ten frames chosen from a video sequence. This is called automatic peak detection and is explained as follows.

Emotion expression can be divided into three parts: outset, apex, offset. In our approach, apex of the emotion has been detected automatically. We have a sequence of 80 frames; we will re-sample this sequence taking only 10 out of 80 frames.

Out of the ten frames chosen, to detect the peak, the Euclidian distance between the corresponding points on neutral frame and the next frames has been calculated. To improve computational time, we will consider only 10 frames from the whole video sequence. We will compute the sum S_k of Euclidean distances between the corre-sponding points on the first frame and frame k to form vector V.

The Euclidean distance between the corresponding points between the first frame and frame k, with n the number of points considered is given in Eq. (1).

$$Sk = \sum_{i=1}^{n} \text{dist}(P_i^0, P_i^k) \tag{1}$$

The vector of Euclidean distances between the first frame and the other frames, with NF the number of frames considered is given in Eq. (2).

$$V = [S_1, S_2, \ldots, S_k, \ldots \ldots, S_{NF}] \tag{2}$$

Index of the maximum of the vector V will correspond to the index of the peak frame.

In most of the papers dealing with dynamic emotion recognition, instead of using neutral and apex frame they have chosen an n-frame window. This doesn't require any peak detection and can be used for real time detection.

3.2 Feature Extraction

After pre-processing steps, as all faces are correctly aligned, and centered in the nose tip (0, 0, 0), the curves are used to re-sample the faces to extract feature information. Two different approaches are presented to re-sample the faces in an intelligent way using parallel curves and radial curves.

3.2.1 Feature Extraction Using Curves

a. Parallel curves

In this approach, profile curves have been used to resample the face and to extract the feature information from these curves to form feature vector. The profile curves extracted on a sample face are shown in the Fig. 2.

Fig. 2. Profile curves

The movement of the face is measured through these curves and for that, two frames are used: neutral frame and the peak frame. The deformation matrix is formed by computing the distances between point to point on the corresponding curves of neutral frame and peak frame. Then, this matrix is used to create the feature vector that is used for classification. In the next paragraph, curve extraction and re-sampling of these curves using different parameters will be presented. Faces are represented by a set of curves evenly spaced. Three parameters can be changed to proceed with curve extraction: (1) Distance (2) Number of curves (3) Number of points.

Distance: This parameter determines the width of the face considered as shown in Fig. 3.

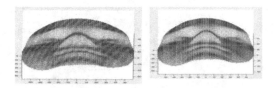

Fig. 3. Influence of the parameter distance for the same number of curves. Left distance = 40, right distance = 80.

Number of curves: This parameter will determine the number of curves taken on a facial surface as shown in Fig. 4.

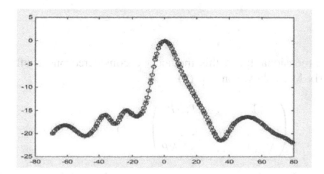

Fig. 4. Different number of curves for distance = 60. Number of curves from Left to right: 7, 11, 17

Number of points: After extraction of this profile sets, the curves have been re-sampled using spline as shown in Fig. 5.

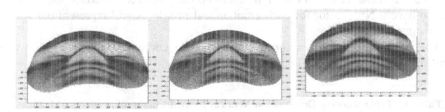

Fig. 5. Example of re-sampling one curve. Blue circles represent the original points, red stars after re-sampling with a chosen number of points. (Color figure online)

After choosing this different parameters and extracting curves, each face will be represented by a set of curves. To measure the face motion the distances between point to point on corresponding curves of neutral frame and peak frame are calculated as shown in Fig. 6.

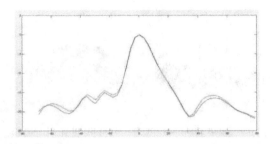

Fig. 6. Comparison between neutral and peak curve after re-sampling

A deformation matrix M for each profile, each emotion is formed. Each coefficient of this matrix represents the distance peak-neutral for one point of one curve.

To use a classifier like SVM or Random Forests we need to reduce the data. We need to have a N*1 vector for each emotion and profile. N will represent our feature vector dimensionality.

The deformation matrices formed using profile curves as features is shown in Table 1.

Table 1. Table showing Deformation Matrices formed using profile curves as features.

What we have now (for each emotion, each profile)	$M = \begin{pmatrix} 1,1 & \cdots & 1,NP \\ \vdots & \ddots & \vdots \\ NC,1 & \cdots & NC,NP \end{pmatrix}$ where NC corresponds to number of curves and NP to the number of points
What we need (for each emotion, each profile)	$V = (1, \ldots\ldots\ldots\ldots\ldots, N)$ where N corresponds to feature vector dimensionality

To reduce dimensionality, in this method we considered one coefficient for each line of the matrix M as shown in Eq. (3)

$$\begin{pmatrix} 1,1 & \cdots & 1,NP \\ \vdots & \ddots & \vdots \\ NC,1 & \cdots & NC,NP \end{pmatrix} \rightarrow (1 \cdots \cdots NC) \qquad (3)$$

Unfortunately this method does not manage to show differences between emotion and we get an average of 60% of recognition for six emotions.

Then, to give more information to our feature vector, we used the whole matrix after transformation as shown in Eq. (4). New feature vector dimensionality will be NC*NP.

$$\begin{pmatrix} 1,1 & \cdots & 1,NP \\ \vdots & \ddots & \vdots \\ NC,1 & \cdots & NC,NP \end{pmatrix} \rightarrow (1 \cdots \cdots NP \cdots \cdots 2*NP \cdots \cdots NP*NC) \quad (4)$$

Using this method a maximum of 76.7% recognition rate has been obtained using specific parameters: NC = 7, NP = 25, distance = 90.

b. Radial Curves

The algorithm is similar to the extraction of parallel curves, but the way the curves are extracted is different. We have performed the same pre-processing steps. Then, we divide the face in different sections choosing an angle as it is shown below in Fig. 7.

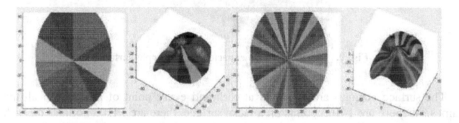

Fig. 7. Two examples of radial curve extraction for different number of angle scale 30°, 10°

To extract curve from each part of the face, each part of the face is rotated by a certain angle as shown in the Fig. 8. and by taking points with x = 0, a curve is obtained corresponding to this particular part of the face.

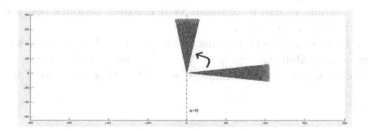

Fig. 8. After extraction, all parts of the face are rotated and a curve is extracted.

After that, a test has been conducted: first by taking one value for each curve and then by taking a chosen number of points. Using one value for each curve, better accuracy than using profile curves has been obtained, (70% with angle step = 5). Using most important number of points for each curve, better accuracy is obtained (81.7% recognition for the six emotions).

Thus, using radial curves maximum accuracy of 81.7% has been obtained. How-ever this method is an off-line method because it needs peak frame.

3.2.2 Feature Extraction Using Surface-Normals

Surface-normals can be used as 3D features. At each vertex in the facial surface, normals are extracted and each point on a facial surface can be encoded using it's normal vector.

$F_i^N = \left\{ n_1^i, n_2^i, \ldots \ldots, n_m^i \right\}$, where F_i is the face of an ith individual, $n_j^{i'}$ s are 3D unit normals: $n_j^i = \{n_x, n_y, n_z\}$. The surface normals obtained for a sample surface is shown in the Fig. 9.

Fig. 9. Surface normals extracted for a sample surface

The surface normals are extracted at each and every point of a frame for all the frames. As there are 14103 points on a facial surface, there are 14103 normals on each and every frame. The feature vector is formed by calculating the euclidean distances between the corresponding normals of neutral frame and peak frame.

3.3 Classification

For classification, Support Vector Machine (SVM) is used which is common in machine learning and pattern recognition. SVM classifier performs best and is used generally to classify binary classes.

Basically SVM is used to classify data containing two classes. In our work, as we have six emotions, Multiclass Support Vector Machine (MSVM) has been used. Figure 10 an example is shown illustrating how SVM classifies two sets of points using a plane.

Fig. 10. SVM perform for two sets of points

This example illustrate perfectly what SVM is doing: it tries to separate the two sets of points of dimension (n) using a hyper plane of dimension (n−1) which can classify with the maximum margin. In this case, as there are a set of points of dimension 2, the two set of points are separated by using a hyper plane of dimension one. This example shows how SVM works using linear kernel, but other kernels like Gaussian or polynomial can also be used.

4 Experiments and Results

The proposed algorithms are tested on the 60 subjects of BU-4DFE Database that have the 3D video sequences of six basic emotions of each subject. Out of the 60 feature vectors which are formed for 60 subjects, 40 feature vectors are randomly selected for training and 4 samples of 5 subjects are used for testing. The classification is done by using SVM classifier and the results obtained are shown in Tables (2), (3) and (4) respectively. 7 profile curves are taken on each facial surface and on each curve 25 points are taken. In the radial curve based approach, 120 curves are taken on each facial surface and 10 points on each curve. In the surface normal based approach, 14103 normals are extracted on each facial surface. The average recognition rate is 76.6%, 81%, 80% for parallel curves, radial curves and surface normals respectively. Happy and surprise expressions have the best recognition rates. The recognition rates are given in percentage.

Table 2. Confusion matrix using parallel curves

	Anger	Disgust	Fear	Happy	Sad	Surprise
Anger	75	0	0	0	25	0
Disgust	10	75	10	0	0	5
Fear	0	15	60	15	0	10
Happy	0	0	15	85	0	0
Sad	15	0	0	0	75	10
Surprise	0	0	10	0	0	90

Table 3. Confusion matrix using radial curves

	Anger	Disgust	Fear	Happy	Sad	Surprise
Anger	85	5	0	0	10	0
Disgust	15	75	5	0	0	5
Fear	0	10	70	20	0	0
Happy	0	0	10	90	0	0
Sad	20	0	0	0	75	5
Surprise	0	0	0	5	0	95

Table 4. Confusion matrix using surface normals

	Anger	Disgust	Fear	Happy	Sad	Surprise
Anger	70	0	0	5	25	0
Disgust	15	75	0	0	10	0
Fear	5	5	70	5	10	5
Happy	0	5	5	85	5	0
Sad	25	0	0	5	70	0
Surprise	0	0	0	0	15	85

The results are compared with existing literature and is shown in Table 5. The recognition rate obtained by Georgia [11], based on temporal analysis was 80.04%. In this method, subsequences of frames of constant window width equal to six have been taken, by avoiding peak frame detection and thus can be used for real time applications. But, in this method 83 manually annotated landmarks are required on the first frame of the sequence to allow accurate model tracking, hence requiring human intervention. In [12], a method had been proposed using 3D motion based features between frames of 3D video sequences to perform dynamic facial expression recognition. In this method, feature selection methods are applied to each of the onset and offset segments of the expression. An average recognition rate of about 73.61% (per frame) was achieved by using this approach. In [7], annotated landmarks are manually extracted requiring human intervention.

Table 5. Results comparison

Reference	Recognition rate	Remarks
11	80.04%	Subsequences of frames of constant window width equal to six have been taken, by avoiding peak frame detection. In this method, 83 manually annotated landmarks are required on the first frame of the sequence to allow accurate model tracking, requiring human intervention
12	73.61%	3D motion based features between frames of 3D video sequences are used for performing dynamic facial expression recognition
Feature extraction using profile curves	76.6%	This method requires peak frame detection
Feature extraction using radial curves	81%	By using radial curves good recognition rate has been achieved and this method also requires peak frame detection
Feature extraction using surface normals	80%	Good recognition rate has been achieved using surface normals and the algorithm for extraction of normals is not complex

5 Conclusion and Future Work

In this work six basic emotions have been recognized using the curve based approach and surface normal based approach. The proposed algorithm followed for the extraction of surface normal is simple and the recognition rate achieved by using surface normal is acceptable. The preprocessing steps show 100% nose detection on the 60 subjects tested. But, the complexity of the algorithm is high, so it should be reduced to decrease the computational time.

The proposed algorithms are based on peak detection, and this makes them not suitable in real time. By choosing one of the three proposed approaches the algorithm has to be improved to suit real time needs. In future, human-machine interaction would have many applications. By modifying this work to real time emotion recognition, many interesting applications can be developed. The different areas that need to be improved are pre-processing and peak detection.

References

1. Drira, H., Amor, B.B., Daoudi, M., Srivastava, A., Beretti, S.: 3D dynamic expression recognition based on a novel deformation vector field and random forest. In: Pattern Recognition 21st International Conference, pp. 1104–1107 (2012)
2. Gokberk, B., M, Okan Irfanoglu, Akarun, B., Gokberk, L.: 3D shape-based face representation and feature extraction for face recognition. Image Vis. Comput. **24**, 857–869 (2006). ELSEVIER
3. Amor, B.B., Drira, H., Beretti, S., Daoudi, M., Srivastava, A.: 4D facial expression recognition by learning geometric deformation. IEEE Trans. Cybern. **44**, 2443–2457 (2014)
4. Le, V., Tang, H., Thomas Huang S.: Expression recognition from 3D dynamic faces using robust spatio-temporal shape features. In: IEEE International Conference on Automatic Face and Gesture Recognition and Workshops, pp. 414–421 (2011)
5. Gokberk, B., Salah, A.A., Akarun, L.: Rank-based decision fusion for 3D shape-based face recognition. In: Proceedings of the IEEE 13th Signal Processing and Communications Applications Conference, pp. 364–367 (2005)
6. Suja, P., Kalyan Kumar, V. P., Shikha T.: Dynamic facial emotion recognition from 4D video sequences. In: Conference on Contemporary Computing (IC3 2015), pp. 348–353 (2015)
7. Kalyan Kumar, V.P., Suja, P., Tripathi, S.: Emotion recognition from facial expressions for 4d videos using geometric approach. In: Thampi, S.M., Bandyopadhyay, S., Krishnan, S., Li, K.-C., Mosin, S., Ma, M. (eds.) Advances in Signal Processing and Intelligent Recognition Systems. AISC, vol. 425, pp. 3–14. Springer, Heidelberg (2016). doi:10.1007/978-3-319-28658-7_1
8. Sun, Y., Yin, L.: Facial expression recognition based on 3d dynamic range model sequences. In: Proceedings of the 10th European Conference on Computer Vision: Part II, ECCV, pp. 58–71, 2008
9. Zhang, X., Yin, L, Cohn, J.F., Canavan, S., Reale, M., Horowitz, A., Liu, P.: A high resolution Spontaneous 3D dynamic facial expression database. In: IEEE International Conference and Workshops on Automatic Face and Gesture Recognition, pp. 1–6 (2013)

10. Fang, T., Zhao, X., Ocegueda, O., Kakadiaris, I.A.: 3D facial expression recognition: a perspective on promises and challenges. In: IEEE International Conference on Automatic Face and Gesture Recognition and Workshops, pp. 603–610 (2011)
11. Sandbach, G., Zafeiriou, S., Pantic, M., Rueckert, D.: A dynamic approach to the recognition of 3D facial expressions and their temporal models. In: IEEE International Conference on Automatic Face and Gesture Recognition and Workshops, pp. 406–413 (2011)
12. Pantic, M., Patras, I.: Dynamics of facial expression: recognition of facial actions and their temporal segments from face. IEEE Trans. Syst. Man Cybern. Part B 36, 443–449 (2006)

Brain Machine Interaction

Classification of Indian Classical Dance Forms

Shubhangi[(⊠)] and Uma Shanker Tiwary[(⊠)]

Department of Information Technology, Indian Institute of Information
Technology, Allahabad, India
shubz247@gmail.com, ust@iiita.ac.in

Abstract. The algorithm proposed in this paper aims to achieve pose recognition in Indian classical dance domain. Three different dance forms namely Bharatnatyam, Kathak and Odissi, all together with 15 poses have been considered for pose classification problem. An initial database is created consisting of 100 images and split into training and testing dataset. Hu moments have been chosen as the feature extraction technique to describe the shape context of an image since they are scale, translation and rotation invariant. To extract Hu moments from the image, the foreground and the background of the images must be separated. The resultant images are then converted to binary. Since it is a multiclass classification problem, SVM using one vs one approach as well as one vs all approach has been implemented and the results are contrasted with linear and RBF kernels for both the approaches.

Keywords: Pose recognition · Hu moments · Multiclass SVM · Grabcut algorithm

1 Introduction

India has a thousand year old tradition of fine arts and classical and folk music and dances. Bharatnatyam, Kathak, Kathakali, Manipuri, Kuchipudi, Mohiniattam and Odissi are some of the dance forms that originated and evolved in India. The dance forms studied here are Odissi, Kathak and Bharathnatyam. One of the most prominent features of Odissi are **Bhangas** or stance, which involves stamping of the foot and striking various postures as seen in Indian sculptures. The common **Bhangas** are *Bhanga, Abanga, Atibhanga* and Tribhanga [1]. Bharathnatyam is known for its grace, elegance, purity, tenderness, expression and sculpturesque poses. Adavus forms the basic of bharathanatyam. Adavus are combined to form a dance sequence in Bharatanatyam. An adavu is the combination of position of the legs (Sthanakam), posture of our standing (Mandalam), walking movement (Chari) and hand gestures (nritta hastas) [2]. Kathak is the major classical dance form of northern India. Meaning of the word kathak is "to tell a story". A short dance composition is known as a tukra. Most compositions have 'bols' or rhythmic words which serve as mnemonics to the composition [3]. A dance composition consists of mainly tatkar (footwork), hastaks (hand movements), chakkars (circular movements) etc. The proposed algorithm aims to classify images into different poses belonging to Bharatnatyam, Odissi and Kathak. This paper aims to apply computer vision and image processing techniques beyond

A. Basu et al. (Eds.): IHCI 2016, LNCS 10127, pp. 67–80, 2017.
DOI: 10.1007/978-3-319-52503-7_6

simple pose recognition such as walking, running, jumping etc. An application of the proposed technique can be extended to learn basic poses of the three dance forms.

Indian classical dances consist of various body poses, gestures and mudras peculiar to each dance. Depending on the pose we can classify the images into different dance forms. In general Indian classical dance classification is an application of pose recognition. There have been a few attempts to classify dance poses or recognize them. In [4] the authors aim to classify the dances into Bharathanatyam, Odissi and Kathak from dance videos. Each frame of a dance video is represented by a pose descriptor which is based on histogram of oriented optical flow in a hierarchical manner. Online dictionary learning technique is used to learn pose basis. Each video is then represented sparsely as a dance descriptor by pooling pose descriptor of all the frames. For classification, SVM with intersection kernel has been used. Gesture recognition algorithm for Indian Classical Dance gesture recognition using Kinect sensor has been proposed by authors of [5]. A skeleton of human body is created using Kinect sensor from which twenty different junction coordinates are obtained. They have implemented a system of feature extraction which can distinguish between 'Sadness', 'Anger', 'Happiness', 'Fear', and 'Relaxation'. With the help of its intensity information the system checks whether the emotion is positive or negative. Authors of [6] have proposed a technique for extracting the features from the relational graph obtained from the articulated upper body poses for Bharatanatyam dance steps. Features are extracted from an attributed relational graph from images with clothing diversity, background clutters and illumination. Skeletonization process forms the first part of the proposed algorithm which determines the human pose and the smoothness is increased using B-Spline approach. From the generated attributed relational graph geometrical features are extracted so that the correct shape can be recognized and classified into the corresponding pose. In [14] the authors have proposed an algorithm for estimating dancing skills from videos based on rhythmical factors. Firstly, silhouette from the images of dancers is extracted using chroma-key technique, subsequently body parts are detected using skin color information. Once the hand and head parts are detected they are parameterized by representing them with geometrical parameters. This is followed by morphological skeletonization on these body parts. Hough transformation is carried out on these skeletons to represent the generated motion parameters in a simple form. The motion parameters are then tracked to analyze the correlation between the parameter set and dance steps. In [16] the authors aim to detect mudras of Bharathnatayam dance form being performed by the subject. To distinguish the object from background and to get salient features of the static mudra image, hypercomplex representation of the image is taken. For classification purpose K nearest neighbor algorithm is used. A robust feature extraction technique is trivial for any classification problem to be effective, it should be scale, translation a rotation invariant. For this purpose Hu moments have been employed [7]. Handwritten digits recognition represents other branch of this research field. Handwritten digits recognition is an application of shape recognition and classification. An algorithm for handwritten digits recognition based on Hu moments has been proposed by the authors of [13]. Handwritten digits can be of varied shapes which makes template based recognition is difficult to implement. Hu moments are invariant to transformations like rotation, scaling and translation and the proposed algorithm achieves acceptable recognition ratios on standard MNIST database. Hu moments have been applied by various authors for object recognition and as shape descriptor for human action recognition but to the best of my knowledge they have not been applied for pose recognition in the dance domain. SVM has been used for classification purpose in the algorithm proposed in this paper. The authors of [9] have compared the performance of one vs one SVM [10] and one vs all SVM classification [11] techniques that contradict each other on the use of appropriate SVM approach. As stated by researchers the technique apt for the algorithm depends on the type of data and the features selected for classification purpose. The rest of the paper is organized as: Proposed methodology and results are described in Sects. 4 and 5 respectively and finally Sect. 6 concludes the paper.

2 Dataset Description

The dataset comprises of 100 images which have been divided into training and testing dataset. Training dataset consists of 70 images and testing dataset consists of 30 images. A total of 7 subjects have been used for dataset acquisition. The dance forms considered for dataset acquisition are Bharathnatyam, Odissi and Kathak. A prerequisite requirement for image acquisition is that the subject's clothes must be contrasting to the background and should not be too loose so that shape context of the body can be calculated accurately. To accomplish the above stated requirement, white background with darker shades of clothes worn by subjects, is maintained throughout the data acquisition process. A total of 15 poses have been considered which are mentioned in Table 1.

Figure 1 shows three different subjects standing in three different pose belonging to Bharatnatyam, Odissi and Kathak respectively.

Table 1. Pose class

Pose	Class
Tatta adavu	Bharatnatyam
Murka adavu	Bharatnatyam
Natta adavu	Bharatnatyam
Tatti metti adavu	Bharatnatyam
Kudittametta adavu	Bharatnatyam
Teermanam	Bharatnatyam
Abhanga	Odissi
Samabhanga	Odissi
Tribhanga	Odissi
Chowkh	Odissi
Hastak 1	Kathak
Hastak 2	Kathak
Hastak 3	Kathak
Chakkar	Kathak
Kran	Kathak

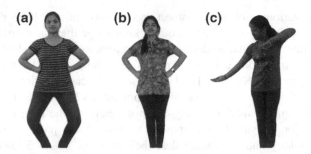

Fig. 1. (a) Tatta adavu pose (b) Samabhanga pose of odissi (c) hastak 2 of kathak of bharathnatyam

3 Experimental Setup

The images were taken with Nikon p600 camera without a tripod, at varied angles and under different lighting conditions to check the robustness of algorithm. Since the images were acquired during different seasons, therefore the clothing of the subjects has changed accordingly.

4 Proposed Methodology

In our method, the foreground and the background pixels are first separated using GrabCut algorithm. Resultant output image contains human figure over a white background. This is followed by binarization of the image, where white region represents human figure and black region represents background. Hu's seven moments are extracted which serve as feature vector and subsequently multiclass SVM is trained with the feature vectors thus obtained. Figure 2 shows the flow of proposed methodology which is discussed in detail in the following sub-sections.

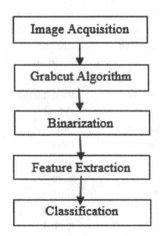

Fig. 2. Architecture of propose methodology

4.1 Image Acquisition

As mentioned in Sects. 2 and 3, first stage of the algorithm is image acquisition.

Figure 3 shows the subject standing in the pose "Samabhanga" belonging to odissi dance form.

Fig. 3. Samabhanga (odissi pose)

4.2 Grabcut Algorithm

As proposed in [8] initially the user is required to draw a rectangle around the foreground region shown in Fig. 4a. The pixels lying outside the rectangle surely belong to background and the pixels lying inside the rectangle may or may not belong to foreground. The algorithm then segments it iteratively to give the desired results, for example if after the initial step some part of background gets extracted along with foreground, the user will explicitly mention which pixels should be labelled as foreground by labeling them as white and as background by labelling them as black and this will continue until the user gets the desire results as depicted in Fig. 4b. A Gaussian Mixture Model (GMM) is implemented to model the foreground and background. Once a model is generated, conditional probabilities can be computed for color pixels. It learns and creates new pixel distribution. This pixel distribution is then used to build a graph, where the nodes in the graph represent pixels. In addition to these nodes, two more nodes are added to this graph, source node and sink node. Every foreground pixel is connected to source node and every background pixel is connected to sink node. The connection between the pixels and source node or sink node is weighted. The weights define the probability of a pixel being a foreground pixel or a background pixel. Pixel similarity is defined by this weight i.e. large weight signifies more similarity in pixel color If there is a large difference in pixel color, the edge between them will get a low weight. The graph is then segmented using a mincut algorithm. It cuts the graph into two separating source node and sink node in such a way that the cost of cutting is minimum. The cost function is the sum of all weights of the edges that are cut. After the graph is cut, all the pixels connected to the source node become foreground pixels and the ones connected to sink node become background pixels. The process is continued until the classification converges i.e. when all the pixels which are supposed to be foreground pixels have been labelled as foreground and the pixels which are supposed to be background labelled as background pixels are labelled as background.

(a) **(b)** **(c)**

Fig. 4. (a) A rectangle is drawn by the by the user (black color), (b) Background pixels are explicity labelled as 0 user around the foreground region, (c) Output image

4.3 Binarization

Hu moments are being used for feature extraction technique and for this technique to work the image must be binary. To obtain a binary image, a suitable threshold is chosen and pixel values above this threshold are set to 1 and below it are set to 0.

Figure 5 is obtained as output when Fig. 4c is converted to binary.

Fig. 5. Binary output image

4.4 Feature Extraction

In any classification problem the efficiency of classification depends heavily on the choice of features. In this classification problem, the feature vectors must be such that it is invariant to scaling, translation and rotation, since images of the subjects posing for the camera are of different size, shape, orientation and relative position. For this reason, Hu's seven moments are chosen as a feature extraction technique [7]. The definition of moments of the gray value-function $f(x, y)$, a two variable function of an object is given as follows:

$$M_{0,0} = \sum\sum x_p y_q f(x, y) dxdy \qquad (1)$$

Since we are using binary images so the gray value function $f(x,y)$ becomes:

$$f(x, y) = b(x, y) = \begin{cases} 1 \text{ Object} \\ 0 \text{ Background} \end{cases}$$

The sum $p + q$ of the indices is the order of the moment m_{pq}.
Considering this, the following moments can be defined as:

- Zero order moment $((p,q) = (0, 0))$

$$m_{0,0} = \sum\sum dxdyb(x, y) \qquad (2)$$

The zero order moment describes the area A of the object.

- First order moments $((p,q) = (1, 0)$ or $(0, 1))$

$$m_{1,0} = \sum\sum dxdyb(x, y) \qquad (3)$$

$$m_{0,1} = \sum\sum dxdyb(x, y) \qquad (4)$$

The first order moments contain information about the center of gravity of the object, where (x_c, y_c) is the centroid of the image.

$$x_c = \frac{m_{1,0}}{m_{0,0}}, y_c = \frac{m_{1,0}}{m_{0,0}} \qquad (5)$$

- Second order moments can be defined as the moment where $((p,q) = (2, 0), (0, 2)$ or $(1,1))$

$$m_{2,0} = \sum\sum x^2 dxdyb(x, y) \qquad (6)$$

$$m_{0,2} = \sum\sum y^2 dxdyb(x, y) \qquad (7)$$

$$m_{1,1} = \sum\sum xydxdyb(x, y) \qquad (8)$$

Second order moments represent the variation of intensity of pixels about origin
Spatial, central and central normalized moments
Since Hu moments feature extraction technique works on binary images, therefore binary images have been used for their explanation in the following section.

- **Invariance to translation**

The centroid of the object is shifted to coincide with the origin of the coordinate system. Using the spatial moments the *central moments* are derived by reducing the spatial moments with the center of gravity (xc, yc) of the object, so all the central moments refer to the center of gravity of the object. Central moments are calculated as follows:

$$\mu_{p,q} = \sum \sum (x - x_c)^p (y - y_c)^q f(x, y) \, dy \, dx \qquad (9)$$

Clearly from the formula,

$$\mu_{0,0} = m_{0,0}$$

$$\mu_{1,0} = \mu_{0,1} = 0 \qquad (10)$$

Figure 6 shows images of chowkh pose(odissi) in which the human figure is at different distances from the origin, therefore translational invariance is required.

Fig. 6. Chowkh pose (odissi) being performed by two different subjects

- *Invariance to scaling*

Scaling invariance is achieved by normalization of each moment. Powers of Area A of the object are used as a scaling factor to get central normalized moments. Figure 7 shows two human figures of different height and width.

Fig. 7. Kran (Kathak) being performed by two different subjects

$$\eta_{pq} = \frac{\mu_{pq}}{\mu_{00^w}} \quad \text{where,} \quad w = \frac{p+q}{2} + 1 \tag{12}$$

η_{pq} is called the normalized central moment.

- *Invariance to rotation*

The seven moments derived by Hu are not only scale and translation invariant but also rotation invariant. They can be derived from central normalized moments as:

$$I_1 = \eta_{20} + \eta_{02} \tag{13}$$

$$I_2 = (\eta_{20} + \eta_{02})^2 + 4\eta_{12} \tag{14}$$

$$I_3 = (\eta_{20} + \eta_{12})^2 + (3\eta_{21} + \eta_{03})^2 \tag{15}$$

$$I_4 = (\eta_{30} + \eta_{12})^2 + (\eta_{21} + \eta_{03})^2 \tag{16}$$

$$I_5 = (\eta_{30} - 3\eta_{12})(\eta_{30} + \eta_{12})[(\eta_{30} + \eta_{12})^2 - 3(\eta_{21} + \eta_{03})] \\ + (3\eta_{21} - \eta_{03})(\eta_{21} + \eta_{03})[3(\eta_{30} + \eta_{12})^2 - (\eta_{21} + \eta_{03})^2] \tag{17}$$

$$I_6 = (\eta_{20} - \eta_{92})[(\eta_{30} + \eta_{12})^2 - (\eta_{21} + \eta_{03})^2] + 4\eta_{11}(\eta_{30} + \eta_{12})(\eta_{21} + \eta_{03}) \tag{18}$$

$$I_7 = (3\eta_{31} - \eta_{03})(\eta_{30} + \eta_{21})[(\eta_{30} + \eta_{12})^2 - 3(\eta_{21} + \eta_{03})^2] \\ - (\eta_{03} - 3\eta_{12})(\eta_{21} + \eta_{03})[3(\eta_{30} + \eta_{12}) - (\eta_{21} + \eta_{03})^2] \tag{19}$$

In Fig. 8 the image on the left is slightly tilted towards left direction, therefore to detect a pose the features selected must be orientation invariant or in other words rotation invariant.

Fig. 8. Tribhanga pose being performed by two different subjects.

The feature vector must be log transformed to compensate for the noise in an image as shown in Fig. 9.

```
File  Edit  Shell  Debug  Options  Windows  Help
Python 2.7.5 (default, May 15 2013, 22:43:36) [MSC v.1500 32 bit (
Type "copyright", "credits" or "license()" for more information.
>>> =============================== RESTART ====================
>>>
array([  2.99111346,    6.53095668,   12.04242492,   10.75174264,
        22.15298964,   14.01748284,   23.00956732])
>>>
|
```

Fig. 9. Feature vector for sample image mentioned in Fig. 5

4.5   Classification

In this work, 6 poses are being considered for bharatnatyam, 4 for odissi and 5 for kathak, making it a total of 15 classes for classification therefore a multiclass classifier is required for this purpose. SVM is inherently a binary classifier but it can be modified to work as multiclass classifier. Two approaches can be adopted for implementing SVM, one vs one and one vs all. One vs all strategy consists of fitting one classifier for each class. For each classifier, the class is fitted against all the other classes. On the other hand one vs one constructs one SVM for each pair of classes, thus it requires c* (c − 1)/2 classifiers for c classes, where each one trains data from two classes. Classification of an unknown data is done according to the maximum voting, where each SVM votes for one class. An advantage of one vs one approach is that it requires less time to train and is preferable for large number of classes. In this project effect of different kernels for one vs one as well as one vs all approach is studied and contrasted. The different kernels considered are linear kernel and rbf kernel. The kernels used in SVM depend on certain parameters such as C and gamma. The mathematical formula for Gaussian kernel is:

$$K(x_i, x_j) = \exp\left(-\gamma \|x_i - x_j\|^2\right), \gamma > 0 \text{ where } \gamma = \frac{1}{2\sigma^2}$$

Since gamma is inversely proportional to σ^2 so a higher gamma means lesser σ^2 and converse, therefore the width of the bell shaped gaussian function is inversely proportional to gamma. If this width is smaller than the minimum pair-wise distance for data, it leads to over fitting and if this width is larger than the maximum pair-wise distance for the data, all points fall into one class. The C parameter establishes inverse relation between the accuracy of classification and simplicity of the decision surface. Figure 9 shows performance of SVM with variation in values of C parameter. Figure 9 (a) denotes a hard margin which tries to find a hyperplane that has maximum distance from either classes' data points, such that all the data points in each side of the hyper-plane should be of the same class, so higher the value of C, more the error is being penalized. Figure 9(c) denotes a softer margin which allows errors to occur while fitting the model (Fig. 10).

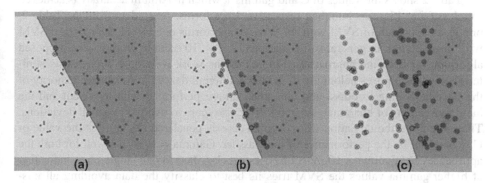

Fig. 10. (a) C = 1000, (b) C = 10, (c) C = 0.1

In this paper the effect of these parameters is studied and the suitable values of these parameters are found for the two approaches using grid search.

5   Results and Discussion

SVM classifies the images into respective pose belonging to one of the three different dance forms. In this paper SVM with following approaches is implemented:

- one vs one approach using linear and rbf kernel
- one vs all approach using linear and rbf kernel. Table 2 summarizes the results obtained for the 2 approaches with both the kernels.

Table 2. Results for different kernels and parameters for inter-dance classification across all poses of the three dance forms

Types of kernels	Accuracy (%)	Parameters
One vs one linear kernel	83.33	C = 3
One vs one rbf kernel	80.0	C = 4
		Gamma = .1
One vs all linear kernel	60.0	C = 5
One vs all rbf kernel	90.0	C = 4
		Gamma = .1

The achieved accuracy for inter-dance pose classification is 90% with a computation time of 15 s on intel i7 processor using one vs all approach of SVM with rbf kernel. The second best accuracy achieved is 83.33% with a computation time of 7 s using one vs one approach with linear kernel. Therefore, one vs all approach proves to be more accurate and expensive in terms of time taken during training process. A tradeoff between time and accuracy exists and the choice of the approach and kernel depends on the kind of application.

Table 2 shows the values of C and gamma at which maximum accuracy is achieved for classifying 15 poses into their respective dance forms. As evident from the Table 2, one vs all approach of SVM with rbf kernel outperforms one vs one approach of SVM with linear kernel. In general softer margins i.e. C < 10 seem to work for the proposed algorithm for both the approaches. Softer margins work well since there exists similarity among the poses belonging to the three different dances, for example majority of the pose belonging to odissi and bharatnatyam are performed with legs bent and the hand movements of bharatnatyam and kathak also seem to overlap by small deviations. Therefore more the deviation from original class is penalized i.e. more is the value of C, worse will be the performance of the classifer. Gamma = .1 works well for both the approaches. At gamma = 10 most of poses get classified to one single class indicating at higher gamma values the SVM tries its best to classify the data avoiding all mis-classifications thus leading to overfitting. A clear distinction between the performance of classifiers i.e. the approach as well as the type of kernel can be made once the database is scaled. At this stage an optimally tuned rbf kernel with one vs rest approach outperforms one vs one approach with linear as well as rbf kernel. Confusion matrix for inter dance classification using rbf kernel and one vs all SVM is given in Table 3.

Table 3. Confusion matrix for inter dance classification

Dance forms	Bharatnatyam	Odissi	Kathak
Bharatnatyam	91.6	8.3	0
Odissi	12.5	87.5	0
Kathak	10	0	90

From the above table it can be seen that 8.3% of Bharatnatyam pose get misclassified as odissi, suggesting that a lot of simililarity exists between Bharatnatyam and Odissi whereas the confusion between Kathak and either of the dance forms is much lower. Table 4 gives accuracy achieved in classifying a pose within a particular dance form.

Table 4. Results for intra-dance classification

	Bharatnatyam	Odissi	Kathak
One vs all SVM with rbf kernel	91.6%	87.5%	80%
One-vs-one SVM with linear kernel	91.6%	87.5%	80%

For intra-dance classification highest accuracy of 91.6% is achieved for bharatnatyam, followed by 87.5% for kathak and 80% for odissi. The performance for one vs one linear kernel and one vs all rbf kernel for intra dance classification is comparable. It can be concluded that the pose belonging to the same dance are linearly separable hence linear kernel seems to work well.

6 Conclusion and Future Work

Indian classical dances are being recognized for the first time using the proposed algorithm. The problem has been handled in the video domain [4] and by employing Kinect sensor [5], but so far Indian classical dance classification from 2D images has not been implemented, moreover this approach doesn't require any special gear, it can be performed using smartphone camera too and is easy to implement.

Future work could include its application to videos by creating temporal templates and using HMM for classification which will enable us to classify the gesture rather than a static pose. Mudras or hand formations have not yet been incorporated in this paper. In the future a hierarchical layer could be built where first the pose are classified into the type of dance say by capturing the shape context of dress code and then subsequently into the type of pose within a particular dance. For example odissi performers wear a crown on their head or tahai whereas bharatnatyam and kathak performers do not and Kathak performers generally wear salwar kameez or lehenga choli. Also some poses can be common for these dances, in that case mudra or hand formations can be used to distinguish among the poses, they can be recognized using fuzzy attributed graphs, salient techniques etc.

References

1. https://en.wikipedia.org/wiki/Odissi
2. http://onlinebharatanatyam.com/2007/07/21/more-about-adavus/
3. https://en.wikipedia.org/wiki/Kathak
4. Samanta, S., Purkait, P., Chanda, B.: Indian classical dance classification by learning dance pose bases. In: 2012 IEEE Workshop on Applications of Computer Vision (WACV). IEEE (2012)
5. Saha, S.: Pose recognition from dance video for Elearning application. Diss. Jadavpur University, Kolkata-700032, India (2011)
6. Sugathan, A., et al.: Attributed relational graph based feature extraction of body poses in indian classical dance bharathanatyam. Int. J. Eng. Res. Appl. 4(5), 11–17 (2014). ISSN 2248-9622, www.ijera.com
7. Ming-Kuei, H.: Visual pattern recognition by moment invariants. IRE Trans. Inf. Theor. 8 (2), 179–187 (1962)
8. Rother, C., Kolmogorov, V., Blake, A.: Grabcut: interactive foreground extraction using iterated graph cuts. ACM Trans. Graph. (TOG) 23(3), 309–314 (2004)
9. Milgram, J., Cheriet, M., Sabourin, R.: "One Against One" or "One Against All": which one is better for handwriting recognition with SVMs? In: Lorette, G. (ed.) Tenth International Workshop on Frontiers in Handwriting Recognition, La Baule (France), Suvisoft, October 2006. Inria-00103955
10. Allwein, E.L., Schapire, R.E., Singer, Y.: Reducing multiclass to binary: a unifying approach for margin classifiers. J. Mach. Learn. Res. 1, 113–141 (2001)
11. Rifkin, R., Klautau, A.: In defense of one-vs-all classification. J. Mach. Learn. Res. 5, 101–141 (2004)

12. Joutsijoki, H., Juhola, M.: Comparing the one-vs-one and one-vs-all methods in benthic macroinvertebrate image classification. In: Perner, P. (ed.) MLDM 2011. LNCS (LNAI), vol. 6871, pp. 399–413. Springer, Heidelberg (2011). doi:10.1007/978-3-642-23199-5_30
13. Zokovich, S., Tuba, M.: Hu moments based handwritten digits recognition algorithm. Ministry of Science, Republic of Serbia, Project No. 44006 (2013)
14. Naemura, M., Suzuki, M.: Extraction of rhythmical factors on dance actions thorough motion analysis. In: Seventh IEEE Workshops on Application of Computer Vision, WACV/MOTIONS 2005, vol. 1. IEEE (2005)
15. Megavannan, V., Agarwal, B., Babu, R.V.: Human action recognition using depth maps. In: 2012 International Conference on Signal Processing and Communications (SPCOM). IEEE (2012)
16. Mozarkar, S., Warnekar, C.S.: Recognizing bharatnatyam mudra using principles of gesture recognition gesture recognition. Int. J. Comput. Sci. Netw. 2(2), 46–52 (2013)

BCI Augmented Text Entry Mechanism for People with Special Needs

Sreeja S.R.$^{(\boxtimes)}$, Vaidic Joshi, Shabnam Samima, Anushri Saha,
Joytirmoy Rabha, Baljeet Singh Cheema, Debasis Samanta, and Pabitra Mitra

Indian Institute of Technology, Kharagpur, India
getsreeja@gmail.com

Abstract. The ability to feel, adapt, reason, remember and communicate makes human a social being. Disabilities limit opportunities and capabilities to socialize. With the recent advancement in brain-computer interface (BCI) technology, researchers are exploring if BCI can be augmented with human computer interaction (HCI) to give a new hope of restoring independence to disabled individuals. This motivates us to lay down our research objective, which is as follows. In this study, we propose to work with a hands-free text entry application based on the brain signals, for the task of communication, where the user can select a letter or word based on the intentions of left or right hand movement. The two major challenges that have been addressed are (i) interacting with only two imagery signals (ii) how a low-quality, noisy EEG signal can be competently processed and classified using novel combination of feature set to make the interface work efficiently. The results of five able-bodied users show that the error rate per minute is significantly reduced and it also illustrates that it can be further used to develop better BCI augmented HCI systems.

Keywords: Human computer interaction (HCI) · Brain computer interface (BCI) · Motor imagery (MI) · People with special needs · Electroencephalogram (EEG) · Text entry system

1 Introduction

About 15% of the world's population, some 785 million people in the globe are suffering from mental and physical disabilities, including about 5% of children, according to a new report prepared jointly by the World Health Organization and the World Bank [3]. Of these, over 5% of the world's population, that is, 360 million people have motor disabilities (328 million adults and 32 million children), where most of the people are from developing and under developed countries [3]. People with motor disabilities experience difficulties coping with the demands that are placed upon them from the environment. They always depend on some individual to communicate and restrict themselves from entertainments and joyful society.

© Springer International Publishing AG 2017
A. Basu et al. (Eds.): IHCI 2016, LNCS 10127, pp. 81–93, 2017.
DOI: 10.1007/978-3-319-52503-7_7

Over years the way humans interact with computers have made remarkable progress, from punch cards to swipe cards to touchless system. Nevertheless, the HCI system designed for disabled lags behind. In fact, for those people with severe disabilities, who can't use their hands and legs properly, only HCI-based systems are not compatible to meet their special needs. If we take a look at the various assistive technologies [6] for disabled users, be it screen readers, eye tracker or something else, we find that all of these use the same data flow path from human brain to hands or some other body part to computer peripherals like camera, keyboard to the computer memory or CPU. What if the humans only think actively and computers somehow understands the users' intention? This forms the underlying capability of BCI where it distinguishes different patterns of brain activity, each being associated to a particular intention or mental task, to directly control the HCI application. Hence, augmenting BCI with HCI and creating hands-free, touch-free applications can greatly enhance the life of people with disabilities.

At present, quite a few functional imaging modalities like EEG, MEG, fMRI, fNIRS, etc. are available for research [16]. Among them the electroencephalography (EEG) is unique and most often used, since it promises to provide high temporal resolution of the measured brain signals, relatively convenient, affordable, safe and easy to use BCI for both healthy users and the disabled. The term EEG is the process of measuring the brain's neural activity as voltage fluctuations along the scalp due to the current flow between neurons. BCI technology has traditionally been unattractive for serious scientific investigation. However, this context undergone radical changes over the last two decades. Now it is a flourishing field with a huge number of active research groups all over the world [16].

The assistive technologies available nowadays lack either speed or accuracy or both. These technologies require interactions that are explicit and exhaustive [6]. Moreover, it is not easy to apply BCI systems to operate an application like virtual keyboard. There are two main challenges associated with this. First, multi-class interaction, that is, more than two gesture usages will reduce the accuracy of the system and make the protocol design complicated [10]. Hence, the system should be designed to work with minimal number of interaction. Second, how a poor quality, noisy EEG signals from an affordable device can be expertly processed to work the interface effectively.

To overcome the above problems, we proposed a BCI augmented text-entry application for people with limited mobility, so that the user can interact with the system using their left and right hand motor imagery brain signals. The proposed approach would be with comparable performance for disabled and able-bodied users, and would not be exhaustive to work with. More specifically our goal is to explore and evaluate various pre-processing methods and feature set, to combine them with classification methods and to finally find a good set of techniques for our project. Besides, users do not have to use any additional equipment like eye tracker, screen readers to improve the system performance.

To accomplish our objective, we carried out our work in five phases. In the first phase, EEG signals are collected from different subjects performing left and right hand motor imagery intentions. In the second phase, preprocessing is done using several methods like independent component analysis (ICA), regularised common spatial pattern (R-CSP), common average reference (CAR), and common spatial pattern (CSP). In the third phase, 14 set of different features has been extracted from the preprocessed signals. In the fourth phase, linear discriminant analysis (LDA) and support vector machine (SVM) classifiers are modelled using training data. In the fifth phase, BCI augmented text-entry application is developed and the output of the classifier is translated into command to control the application. Finally, the accuracy of the proposed system is evaluated.

2 Literature Survey

Farewell *et al.* [7] describe the development and testing of a P300 component based system for one to communicate through a computer. The system is designed for users with motor disabilities. The alphabets are displayed on a computer screen which serves as the keyboard or prosthetic device. The subject focuses on characters he wishes to communicate. The computer detects the chosen character on-line and in real time, by repeatedly flashing rows and columns of the matrix. The data rate of 0.20 bits/sec were achieved in this experiment. BCI system such as SSVEP or P300 [8] involves visual stimulations to control the interface. It uses flickering LEDs and continuous use of this may cause eye fatigue, epileptic seizures and visual impairment.

Scherer *et al.* [14] use a specially designed GUI and employed three motor imagery signals to control the interface. One MI signal for scrolling the letters and the other two to decide a target letter. Here, the typing speed attained was about 4 cpm. Bin *et al.* [5] aim to improve the low communication speed of the EEG based BCI systems based on code modulation of visual evoked potentials (c-VEP). The target stimuli were modulated by a time-shifted binary pseudo-random sequence. The on-line system achieved an average information transfer rate (ITR) of 108±12 bits/min on five subjects and with a maximum ITR of 123 bits/min for a single subject.

Changes in μ (8–12 Hz) and β (18–25 Hz) rhythms are associated with motor imagery signals [4]. Prasad *et al.* [11] aim at goal-directed rehabilitation tasks leads to enhanced functional recovery of paralyzed limbs among stroke sufferers. It is based on motor imagery (MI) and is also an EEG-based BCI. The MI activity is used to devise neuro-feedback for the BCI user to help him/her focus better on the task. The positive gains in outcome measures demonstrated the potential and feasibility of using BCI for post-stroke rehabilitation.

Ryan *et al.* [13] compared a conventional P300 speller brain-computer interface with a predictive spelling program. Time to complete the task in the predictive speller (PS) condition was 12 min 43 s as compared to 20 min 20 s in the non-predictive speller (NS) condition. Eventhough there is marked improvement in overall output, accuracy was significantly higher in the NS speller. These results demonstrate the potential efficacy of predictive spelling in the context of BCI. Multi-degree BCI system [10] can be used to control various selecting tasks of the user interface like left, right, up, down, backwards, close, delete, open etc. But the accuracy of the classification progressively reduces when the number of classes' increases and more mental task adds complexity in the protocol design.

3   Design of Experiment

In spite of a number of research initiatives, there is still a scope of improving the research direction. This motivates us to lay down our research objective, how only two MI signals can be adopted to develop a low-cost but accurate enough text entry mechanism. As a first step towards this research investigation, we plan to collect training data. Our design of experiment to collect training data is discussed in this section.

3.1   Objective of the Experiment

Our objective is to work with a text-entry application based on neural activity, without any voluntary muscle movements for motor disable people. This hands-free and touch-free system involves typing a letter/word using two simple gestures namely right and left hand MI. The interface consists of specially designed virtual keyboard, where a particular letter is selected by performing MI and dwell time. The commands issued by the user will be displayed in the monitor using visual animations for the quick access and thereby to reduce the decision-making time.

This application system is proposed to assist motor impaired people with good conceptual abilities. There are two types of motor disabilities, traumatic injuries and congenital conditions. Traumatic injuries includes spinal cord injury and loss or damage of limbs. Cerebral palsy, muscular dystrophy, multiple sclerosis, spinal bifida, amyotrophic lateral sclerosis, arthritis and Parkinson's disease comes under congenital conditions [6]. The target users, for our application are assumed to be suffering from such type of disabilities. Severe motor-impaired people, aged disabled people, or too young disabled people are not considered.

3.2   Experimental Setup

Hardware: The Emotiv EPOC+ [2] headset is placed over the scalp according to International 10–20 electrode placement system. It consists of 16 electrodes - 14 measuring electrodes and two reference electrodes. The EEG measuring sensors are AF3, AF4, F3, F4, F7, F8, FC5, FC6, P7, P8, T7, T8, O1, and O2. The reference sensors are P3 (Common Mode Sense - CMS) and P4 (Driven Right Leg - DRL). The sampling rate is 128 sps, cut-off frequency of low pass filter is 45 Hz and the resolution is 14 bits. The recorded EEG signals are transferred to the computer using wireless USB connector. The device has three types of control such as EEG, EMG and Gyroscope. It has fewer scalp contact than a expensive and sophisticated device.

Software: OpenViBE [12] initially stood for Open Virtual Brain Environment, but since it is no longer limited to 'Virtual Environments', OpenViBE is no longer an acronym. OpenViBE is a software platform for Brain-Computer Interfaces (BCIs). It is a free and open source software with capabilities to acquire, filter, process, classify and visualize brain signals in real time. OpenViBE bundle contains various useful tools, and two of the tools namely OpenViBE Acquisition Server and OpenViBE Designer have been utilized in this project.

Participants: Three healthy subjects (two male), right-handed, between ages of 23 and 29, volunteered for experiment. Each subject provided written informed consent before participating in the experiment. None of the subjects had prior BCI training. The subjects were provided with information only related to the activities they performed. They were not informed about the experimental design and the hypothesis/objective of the study. The experimental setup was done at the BCI LAB facility at IIT Kharagpur.

Subjects were seated in chair with their arms extended, resting on the desk and their legs extended, resting on a footrest. The lab was well illuminated by artificial lights, and there was no background noise while the data was recorded. The subjects performed or kinesthetically imagined left and right hand movements, and the EEG signals were recorded on all 16 channels using the Emotiv EPOC+ device and OpenViBE. The subjects were provided with stimulus on GUI (Fig. 1) designed using the OpenViBE tool. They were provided with two stimulus, a cross figure that marked the beginning of the trial followed by a left or a right arrow on which the task was performed, recording EEG signals for MI for nearly 4 s. The stimulus was presented on a nineteen-inch LCD monitor, kept in-front of the subject making a viewing angle of approximately 1.5°. The subject was focused on the stimulus during the entire session. The experiments were performed in presence of a researcher, with no interaction with the subject during the recording session.

3.3   Data Acquisition and Data Labeling

The Data was acquired from each subject over four sessions. Each session lasted about 20 to 30 min, including the time for experimental setup and data recording. The sessions were held on different days. Each session, had 25 trials of left and another 25 trials of right hand movements to be imagined by the subjects. The subjects were shown left and right arrows in random order on the screen and subjects have to think about lifting/moving the corresponding arm. Subjects were instructed to focus on the cues to minimize noisy reading. To minimize the artifacts in the recordings, subjects were asked to minimize eye blinks, jaw and head movements during recording. They were allowed to swallow, blink and adjust to relax during the cross presentation. The duration of cross presentation was adjusted if required for subjects, and also varied randomly.

The data is recorded into a single file as a continuous stream of signals. The data is recorded in random order based on the stimulus the subject received;

Fig. 1. Left and Right arrow shown to subjects to perform corresponding intention.

hence there is no clear demarcation of left and right signals recorded. To ensure proper labeling of stimulus generated, OVTK_GDF_Right for right hand movement, and OVTK_GDF_Left for left hand movement are also recorded alongside to the data file. The data for left versus right movement can be separated using stimulation based epoching box.

4 Proposed Text Entry System

The system can be divided into two sub systems: the BCI sub system and the HCI sub system. Here, the BCI system acts as an interface between the user and the HCI application.

4.1 BCI Subsystem

The BCI sub system involves all the steps of a typical components of BCI system as shown in Fig. 2. The BCI sub system is developed in two steps, First creating a model for the application, that is, off-line analysis, second real-time processing of captured signals, that is, on-line analysis. A large number of signals from the subjects are captured for building a model. Various signal preprocessing techniques are applied to this set of signals, and suitable features are extracted. Generally, the selection of preprocessing method and feature set is an iterative process. An exhaustive set of features are initially selected and applied to the signals, and then each of them is evaluated. The final model will be based on the one with feature vector and preprocessing methods that gave the best accuracy and robustness. Then model is created by training a classifier with the training data and evaluating using cross-validation methods. Once the model is ready, the BCI system is almost ready. Now the signals that are being captured from the subject can be classified in real-time by the same preprocessing techniques and feature vector that were used to train the model. The on-line system will interact with the HCI component to fulfill the subject's intention.

4.2 HCI Subsystem

In our study, we preferred dynamic keyboard [15] for text-entry purpose. This dynamic keyboard ranks good among people with disabilities. This virtual keyboard is designed to have big selection boxes, each box containing letters or

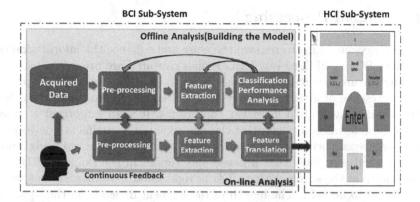

Fig. 2. Radar Curve showing EEG signal of each channel on applying various pre-processing methods.

words that can be selected for typing as shown in Fig. 3. Each box consists of five letters or words arranged with one in top and two in the right and two in the left. This application also consists of the functionality of word prediction to help with spelling. This greatly reduces the search time required to select a word.

To type a letter or word, the user used their MI signals to move the cursor to the target in a circular motion and the corresponding box will be selected if the dwell time is greater than 1.50 s. Once the box is selected, the letters or words are then redistributed among the boxes till the user narrows down to one. Initially, the letters or words appearing in the center of the panel will be automatically selected. If the dwell time is greater than 2.50 s, it is taken as no response from the user and hence the selection box in the center of the panel will get highlighted again. The screenshots of the user typing "I need water" is shown in the above Fig. 3.

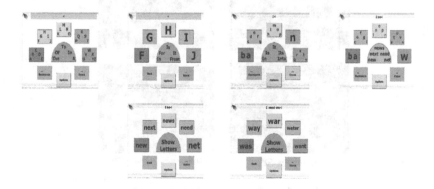

Fig. 3. Screenshots of user typing "I Need Water" using dynamic keyboard.

5   BCI Signal Processing

Signal preprocessing aims to remove the noise and enhance the information content of the EEG signal. The raw EEG signals captured are preprocessed, before performing feature extraction and classification. Another reason for preprocessing is the presence of artifacts, that is, parts of signal due to background brain activity, which may lead to incorrect conclusions. It is an important step as it increases the information content of the raw signal. The artifacts due to power line interference and other artifact above 50 Hz were easily removed using a notch filter, or a band-pass filter that allowed only signals in the range 8–30 Hz. The optimal length of the segments for a left or right hand movement/intent recording was found to be ∼4 s. The length of the segment depends on the signal and kind of features we are looking for.

CAR, ICA, CSP and Regularized CSP were the major preprocessing techniques used and tested for the study. CAR had minor improvements in the results, but use of common average of all channels resulted in artifacts still being present and at some time magnified. ICA, though theoretically was the best method for the preprocessing, is very memory intensive. The requirements of large computation time made it unfit to be used for our project.

In our study, CSP and Regularized CSP gave the best results. CSP and Regularized CSP also reduced the dimensionality from 14 channels data to just 6 channels data as shown in Fig. 4(d). The use of variance between the channels also limits the random artifacts introduced such as eye-blinks. The use of 6 channels gave the best results. Hence, we used Regularized CSP as the preprocessing technique along with the bandpass filter in our project.

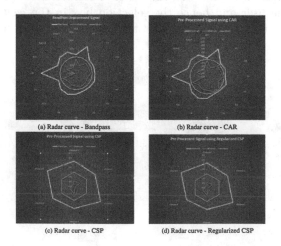

(a) Radar curve - Bandpass   (b) Radar curve - CAR

(c) Radar curve - CSP   (d) Radar curve - Regularized CSP

Fig. 4. Components of BCI system.

6   Feature Extraction and Modeling

Although a large set of features can be extracted, using different methods for BCI component, it is necessary to identify the most efficient features set. The accuracy of the study and the conclusions drawn greatly depends on the features used to represent the entire chunk of the signals for classification. In order to build an effective system, we need to use a small set of features that gives the best results. In our study, to classify the left and the right hand movement/intention, the features listed in Table 1 was extracted.

Table 1. Accuracy of various features extracted using 10-folds cross-validation method on LDA and SVM Classifiers.

Features	Accuracy using SVM in %	Accuracy using LDA in %
Band power of alpha, beta, delta, gamma, mu	58.2	52.5
Statistical features	53.5	50
Valence	46	34
α to β ratio	40	38.5
Signal power [8–30 Hz]	62	56
FFT coefficient	77	72
Hilbert transform	60	53.7
AR With 6 coefficients	60	52
AR With 10 coefficients	55	42
AR With 14 coefficients	65	63
AR With 20 coefficients	85	78
Band power + α to β ratio + Statistical + Valence	57	53
Signal power + Hilbert transform + AR Features	88	70
Signal power + Hilbert transform + AR + FFT	**95**	**87.5**

These extracted features were then evaluated individually and by combining with other features and finally a subset of these features was selected for further processing. The accuracy of these features was evaluated using the k-fold based cross-validation method on LDA and SVM classifiers. From the result obtained it is found that the combination of Hilbert Transform, Auto-regressive Coefficient (20), Band Power (8–30 Hz) and FFT amplitude feature vector produces accuracy of about 95% with SVM classifier and about 87.5% with LDA classifier. Hence this set of feature vector is considered to be the best to train the classifier.

The EEG data collected from the subjects from all sessions were used for classifier training. Classifiers build a model based on the training data, to distinguish between classes. Once the model is created, it could be used to label unseen data i.e. testing data. In order to classify the unseen data correctly the model build must be robust and accurate. This work investigates the performances using Linear Discriminant Analysis (LDA) and Support Vector Machines (SVM) classifiers. These are linear classifiers that use linear functions to differentiate classes. The detailed study of LDA and SVM is found in [9,17]. The result provides the prediction results that directly assigns sample with labels (right hand MI = +1, left hand MI = −1) to identify which category it belongs.

7   Experimental Results and Analysis

The subject is made to work with the BCI augmented HCI based text-entry application using MI signals. The acquired real-time EEG signals will be processed in the same manner as the processing done during building the off-line model. The EEG signal will pass through a band-pass filter to extract EEG in frequency range of 8–30 Hz. The signal is still from 14 channels, this data is pre-processed and reduced to 6 channels data using CSP filter that has been trained off-line previously. The next step is to identify the epochs and classify them. The outcome of the classifier is then translated into control command, which is used to select the target letter or word in text-entry application.

7.1   Evaluation of the Classifier

The correctness of on-line classification with real time EEG data is measured using Confusion matrix and Cohen's kappa coefficient. From confusion matrix, the total-accuracy is calculated using the ratio of sum of true positives (TP) and true negatives (TN) to the total of TP, FP, TN and FN. The Cohen's kappa coefficient assigns equal weight to all classes and it considers the distribution of wrong classification too. From the result shown in Table 2, it is experiential that SVM classifier performs better than LDA classifier in all states.

Table 2. Accuracy and kappa value of LDA and SVM Classifiers.

Classification technique	Accuracy in %	Kappa value
LDA Classifier without preprocessing	45	0.12
LDA Classifier with preprocessing	82	0.64
SVM Classifier without preprocessing	58	0.50
SVM Classifier with preprocessing	**93.5**	**0.89**

7.2 Evaluation of BCI Augmented Text Entry Application

The proposed design is modest and doesn't require any high computational power. The character selection is achieved by staying at the object of interest for a particular amount of time called dwell time. However, the proposed system has one limitation. Increasing the dwell time increases the accuracy of the system but it slows down the speed of the system. The accuracy of the system is evaluated using true positive rate (TPR) and Cohen's kappa coefficient. In our experiment we used the dwell time as 1.50 s. As shown in Table 3, when the dwell time increases from 0.50 s to 2.50 s the average TPR increases from 26.7% to 60.8% and kappa value from 0.37 to 0.66.

Table 3. Comparison of TPR% and Cohen's kappa coefficient on varying dwell time.

Subjects	Dwell time(0.50)	Dwell time (1.50)	Dwell time (2.50)
AA	35.8, 0.50	84.0, 0.88	89.1, 0.80
BB	08.5, 0.14	35.1, 0.48	46.8, 0.59
CC	35.9, 0.48	41.7, 0.54	46.6, 0.59
Average	26.7, 0.37	53.6, 0.63	60.8, 0.66

However, for the practical purpose, typing speed, that is, number of correct letters including space entered by the user in one minute, is the better metric to estimate the performance of BCI augmented text entry system. In our study, the average typing speed attained by the users range from 12 ± 3 cpm. Table 4 shows the comparison of typing speed of three users in character per minute (CPM) on texting "The quick brown fox jumps over the lazy dog" using dynamic keyboard (DK) and microsoft On-Screen keyboard (OSK) [1].

Table 4. BCI augmented Text Entry performance with user evaluation.

Subjects	Text entry rate (Character per minute)	
	OSK	DK
AA	7	14
BB	5	9
CC	9	15
Average (CPM)	7	12.6

8 Conclusion

The proposed BCI augmented text entry system makes use of left and right hand MI signals to operate the text-entry application. This system is intended for people with motor disabilities who have good cognitive capability. The proposed system provides a better alternative to operate a virtual keyboard and it provides true independence to the users without compromising with the accuracy and performance, using the low-priced Emotiv EPOC EEG device. The following observations were made as a part of the study:

- OpenViBE software is comparatively easier to develop an online real-time BCI-HCI application.
- The quality of EEG data collected depends on the environment and the concentration of the subject.
- Regularized CSP is faster and gives better results as compared to preprocessing methods like CAR, ICA.
- The combination of Hilbert transform, AR with 20 coefficients, Band power (8–30 Hz), and FFT amplitude feature vector gives higher precision.
- SVM gives better classification accuracy when compared to LDA.
- Increasing the dwell time increases the accuracy of the system.
- The average typing speed of our system gives better result than using On-Screen keyboard.

Our future work will be focussed on improving the performance of the system by decreasing the dwell time. Hybrid BCIs involving multimodalities can be explored further in future, to form a sophisticated interaction system for people with limited mobility. The typing speed can be increased with multiclass interactions; as a result more sophisticated machine learning algorithms can be employed.

References

1. Microsoft, Type without using the keyboard (On-Screen Keyboard). https://support.microsoft.com/en-in/help/10762/windows-use-on-screen-keyboard
2. Emotiv EPOC, Software Development Kit (2010). http://www.emotiv.com/researchers
3. World report on disability, World Health Organization (2011). http://www.who.int/disabilities/world_report/2011/report/en
4. Alomari, M.H., Samaha, A., AlKamha, K.: Automated classification of L/R hand movement EEG signals using advanced feature extraction and machine learning. Int. J. Adv. Comput. Sci. Appl. 4(6) (2013)
5. Bin, G., Gao, X., Wang, Y., Li, Y., Hong, B., Gao, S.: A high-speed BCI based on code modulation VEP. J. Neural Eng. 8(2), 025015 (2011)
6. Crow, K.L.: Four types of disabilities: their impact on online learning. TechTrends 52(1), 51–55 (2008)
7. Farwell, L.A., Donchin, E.: Talking off the top of your head: toward a mental prosthesis utilizing event-related brain potentials. Electroencephalogr. Clin. Neurophysiol. 70(6), 510–523 (1988)
8. Graimann, B., Allison, B., Pfurtscheller, G.: Brain-computer interfaces: a gentle introduction. In: Graimann, B., Pfurtscheller, G., Allison, B. (eds.) Brain-Computer Interfaces, pp. 1–27. Springer, Heidelberg (2009)

9. Guo, L., Wu, Y., Zhao, L., Cao, T., Yan, W., Shen, X.: Classification of mental task from EEG signals using immune feature weighted support vector machines. IEEE Trans. Magn. **47**(5), 866–869 (2011)
10. Long, J., Li, Y., Wang, H., Yu, T., Pan, J., Li, F.: A hybrid brain computer interface to control the direction and speed of a simulated or real wheelchair. IEEE Trans. Neural Syst. Rehabil. Eng. **20**(5), 720–729 (2012)
11. Prasad, G., Herman, P., Coyle, D., McDonough, S., Crosbie, J.: Using motor imagery based brain-computer interface for post-stroke rehabilitation. In: 2009 4th International IEEE/EMBS Conference on Neural Engineering, pp. 258–262. IEEE (2009)
12. Renard, Y., Lotte, F., Gibert, G., Congedo, M., Maby, E., Delannoy, V., Bertrand, O., Lécuyer, A.: Openvibe: an open-source software platform to design, test, and use brain-computer interfaces in real and virtual environments. Presence **19**(1), 35–53 (2010)
13. Ryan, D.B., Frye, G., Townsend, G., Berry, D., Mesa-G, S., Gates, N.A., Sellers, E.W.: Predictive spelling with a P300-based brain-computer interface: increasing the rate of communication. Int. J. Hum. Comput. Interact. **27**(1), 69–84 (2010)
14. Scherer, R., Muller, G., Neuper, C., Graimann, B., Pfurtscheller, G.: An asynchronously controlled EEG-based virtual keyboard: improvement of the spelling rate. IEEE Trans. Biomed. Eng. **51**(6), 979–984 (2004)
15. Spalteholz, L., Li, K.F., Livingston, N., Hamidi, F.: Keysurf: a character controlled browser for people with physical disabilities. In: Proceedings of the 17th international conference on World Wide Web, pp. 31–40. ACM (2008)
16. Wolpaw, J., Wolpaw, E.W.: Brain-Computer Interfaces: Principles and Practice. OUP USA, New York (2012)
17. Yong, X., Fatourechi, M., Ward, R.K., Birch, G.E.: The design of a point-and-click system by integrating a self-paced brain-computer interface with an eye-tracker. IEEE J. Emerg. Sel. Topics Circ. Syst. **1**(4), 590–602 (2011)

Intelligent Movie Recommender System Using Machine Learning

Abhishek Mahata, Nandini Saini$^{(\boxtimes)}$, Sneha Saharawat,
and Ritu Tiwari

ABV-Indian Institute of Information Technology and Management,
Gwalior, India
iamabhishekmahata@gmail.com,
gettonandinisaini428@gmail.com,
snehasaharawat95@gmail.com, tiwariritu2@gmail.com

Abstract. Recommender systems are a representation of user choices for the purpose of suggesting items to view or purchase. The Intelligent movie recommender system that is proposed combines the concept of Human-Computer Interaction and Machine Learning. The proposed system is a subclass of information filtering system that captures facial feature points as well as emotions of a viewer and suggests them movies accordingly. It recommends movies best suited for users as per their age and gender and also as per the genres they prefer to watch. The recommended movie list is created by the cumulative effect of ratings and reviews given by previous users. A neural network is trained to detect genres of movies like horror, comedy based on the emotions of the user watching the trailer. Thus, proposed system is intelligent as well as secure as a user is verified by comparing his face at the time of login with one stored at the time of registration. The system is implemented by a fully dynamic interface i.e. a website that recommends movies to the user [22].

Keywords: Recommendation systems · Sentimental analysis · Emotion analysis · Age and gender detection · Collaborative filtering · Content-based filtering · Supervised learning

1 Introduction

In today's world with the internet and information explosion associated with it, the consumers are facing the problem of too much choice [8]. To alleviate this issue and help the consumers cope with this problem an intelligent movie recommender system is suggested to guide the viewer and provide them with filtered options. This problem is chosen because of the rapid increase in the interests of people in entertainment field over the web. Also, the system aims to enhance the effectiveness of current movie recommender system by making it independent and training it as per user's interest. The proposed recommender system is hybrid in nature by combining the content-based techniques with collaborative based method to make more accurate recommendations than simple independent approaches. These methods are used to eliminate some of the

© Springer International Publishing AG 2017
A. Basu et al. (Eds.): IHCI 2016, LNCS 10127, pp. 94–110, 2017.
DOI: 10.1007/978-3-319-52503-7_8

commonly witnessed problems in recommender systems such as cold start and the sparsity problem. Cold start is a potential problem which concerns with the issue in which because of the insufficient information about the item, the system cannot draw any inferences [19]. In the case of movie recommender system, it arises for the newly released movies. The movie recommender system makes recommendations by comparing the watching and browsing habits of comparable users (i.e. collaborative filtering) as well as suggests movies that share characteristics with films that a user has rated extremely (content-based filtering).

2 Literature Survey and Research Gaps

A lot of research work is done since the late 90 s on movie recommendation system and sentiment analysis to provide best recommendations to the user. Some recommender systems are based on emotions [2, 13] while other takes into account user's Location [19, 20]. In recent times, a majority of the work is done to implement a hybrid approach in movie recommender system.[2, 8, 9, 17]. The hybrid approach combines the collaborative filtering with context-based filtering to get best of both techniques for better assessing the interests of customers and to overcome the weakness of individual methods.

The primary challenge of movie recommender system based on human emotions is to combine complex movie domain with emotions which are a human interaction domain. Such types of system prove to provide better recommendations to users because it establishes a relation between the emotional states of users and the movies suggested to them. Another approach of providing high-quality recommendation can be achieved by utilizing contextual information such as a location of the user. According to [19] the users who share same geographical location have the similar taste to each other. For instance, users who are located in Florida are mostly interested in watching "Fantasy", "Animation" types of the movie; however, the user from Minnesota State are prefers movies related to "War," "Drama".

In movie recommender system some of the works also focus on implementing collaborative filtering technique [1, 19]. It enhances the accuracy and therefore the quality of recommendations. Some systems take into account user rating to refine the movie list [2, 8, 9, 16]. Algorithms such as K-means are used for recommending movies based on user rating. Movie rating was predicted by using classifier models like Spiking Neural Network (SNN), Multi-layer Perceptron Neural Network (MLP), Decision Tree, and Naive Bayes [17]. In [6] author used two machine learning methods Naïve Bayes Classifier and SVM for sentiment analysis. Another method is to find polarity of each sentiment holding words using Sentiment Classifier [20] (Table 1).

Table 1. Comparison of existing system with proposed system

S. No.	Available system	Proposed system
1.	Collaborative filtering methods that provide recommendations to users based on the similarities between users and items leads to poor recommendation because it does not consider other useful factors such as user's location [1,19]	To eliminate this problem we have combined traditional collaborative filtering with content-based filtering to form a hybrid movie recommendation system
2.	Available recommendations are based on users' perspective, but there are things in the market which are worth concerned and that a user is unaware of. These things that are 'out of the box' must also recommend to the users by the engine [9]	The proposed system recommend movies to the users as per their interest as well as feedback provided by similar kind of users
3.	The existing system works on individual users rating. This may be sometime useless for the users who have different taste from the recommendations shown by the system [9]	The system calculates the similarities between different users and then recommends the movies to them as per the ratings as given by different users of similar tastes. This will provide a precise recommendation to the user
4.	In traditional recommender systems, no contextual information is taken into account for generating recommendations. Besides, user's demand might change with context as well as time which cannot be accomplished through tradition recommender systems [10]	To incorporate user's changing demands, a dynamic movie recommender system that updates regularly has been implemented
5.	Most of the recommended system based on reviews from a user does not involve the weight of neutral reviews [1]	All positive, negative and neutral weights has been calculated for evaluating review rating. Also to optimize the movie rating, user rating has been considered
6.	The other type of existing movie recommended system take user input and predict their preferences [1]	The other type of existing movie recommended system take user input and predict their preferences [1]

3 Methodology

The flow diagram is shown in Fig. 1. A user needs to register him first before watching any movie. While registering, the image of the user is captured and his authenticity is verified. If the user is verified, he/she will be directed to the main site. From the captured image, age and gender of the user is detected. She/he will be recommended a list of movies based on the preferences of users of similar age group and gender and also on the genres he/she prefers to watch on our website. The movie list is ordered by average rating. In the navigation menu, there are many other options available like Latest Movies, Watch History, Genre section etc. Once the user selects a movie to watch the trailer of the movie, his/her face expressions get recorded. The genre of the movie is detected by applying neural network on the recorded emotions. There is also a comment section in which a user can review and rate a movie. Based on the review and rating of all the users, the average rating of the movie is determined.

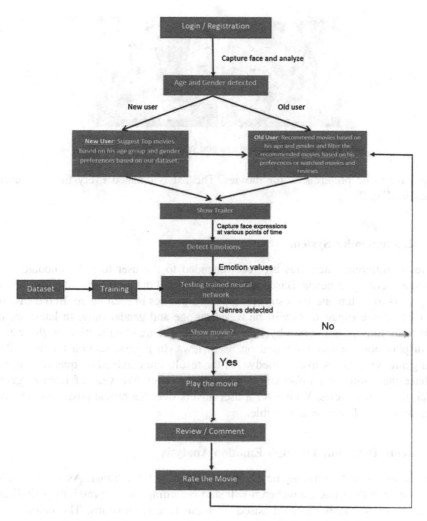

Fig. 1. Framework of movie recommendation system

3.1 Age and Gender Detection

The main objective of this paper is to propose a smart movie recommender system that determines its own input instead of bothering user to provide information. Thus instead of manually taking age and gender, the attributes are obtained from the face image of the user using Face API [5]. The process of estimation of age and gender involves three stages: Pre Processing, Feature Extraction and Classification. At the time of Login, face is captured again and compared with database to verify the user. By capturing a single image of user we are able to detect age and gender as well verifies face. In order to suggest movies to the user, all those users who have the same gender and belong to the same age group as that of the user for whom we need to provide recommendations are considered. The age interval of 10 years is considered as usually people within this age

Fig. 2. Example of age and gender Detection [5]

group have same preferences for movies. The list is updated every time the user is logged in (Fig. 2).

3.2 Recommender System

Movies in different categories are recommended to the user to accommodate user's interest and changing needs. Based on watch history of the user, we evaluate genres the user prefers to watch and then suggest top rated movies in each genre. In the case of a new user, movie suggestions will be based on age and gender only. In latest section, recommend movies are sorted by year of release. The trending section on the website will display popular movies based on user views. In genres, section movies will be listed genre-wise like horror, comedy. All the results are obtained by querying an SQL database that stores the information related to movies like year of release, genres, overall rating and views. Whenever a user hovers over the movie poster movie name, rating and year of release are visible.

3.3 Genre Detection Through Emotion Analysis

After the user selects a movie, he is redirected to watch the trailer. As soon as he/she starts watching the trailer, a webcam will start recording face expressions periodically. The genre of movie is detected based on these face expressions. The emotions are identified using Emotion API [4]. The image is taken as input and confidence across a set of emotions is returned. The emotions detected are happiness, sadness, surprise, anger, fear, contempt, disgust or neutral. All the emotion values which are greater than or equal to 0.9 are taken into consideration. To detect the genre of movie as per viewer's expressions, these emotions are used to train the neural network [3]. A dataset of 500 face samples [18] is constructed for classifying the face expressions to the corresponding genres. Images of faces for all 8 expressions are gathered. A local adaptive learning scheme, RPROP (Resilient Back propagation) performs supervised batch learning in multilayer perceptrons. The magnitude of the weight change is determined by Δ_{ij}^t update-value and direction is determined by sign of the derivative.

Figure 3 represents the summed gradient information over all patterns of the pattern set ('batch learning'). Δ_{ij}^t is calculated on the basis of sign dependent adaptation process as described below: Fig. 4

$$\Delta w_{ij}^{(t)} = \begin{cases} -\Delta_{ij}^{(t)} & , \quad \text{if } \frac{\partial E}{\partial w_{ij}}^{(t)} > 0 \\ +\Delta_{ij}^{(t)} & , \quad \text{if } \frac{\partial E}{\partial w_{ij}}^{(t)} < 0 \\ 0 & , \quad \text{else} \end{cases}$$

Fig. 3. The magnitude of the weight change

$$\Delta_{ij}^{(t)} = \begin{cases} \eta^+ * \Delta_{ij}^{(t-1)} & , \quad \text{if } \frac{\partial E}{\partial w_{ij}}^{(t-1)} * \frac{\partial E}{\partial w_{ij}}^{(t)} > 0 \\ \eta^- * \Delta_{ij}^{(t-1)} & , \quad \text{if } \frac{\partial E}{\partial w_{ij}}^{(t-1)} * \frac{\partial E}{\partial w_{ij}}^{(t)} < 0 \\ \Delta_{ij}^{(t-1)} & , \quad \text{else} \end{cases}$$

Fig. 4. Sign dependent adaptation process

Where $0 < \eta - < 1 < \eta +$

Working of adaptation-rule:

The update value of the weight is multiplied by a factor η , when there's a sign change of the partial derivative of the total error function as compared to the last iteration for each weight or multiplied by a factor of η^+ if there is no sign change. In order to minimize the total error function all the update values are then calculated for individual weight in the above manner, and then finally each and every weight is modified as per its own update value, in the opposite direction of the weight's partial derivative. Based on empirical results, η^+ is set to 1.2 and η^- to 0.5.

The neural network is trained with the input and output as received from the PHP response in the server with RPROP as the "Propagation Trainer". The system has managed to get 96% accuracy. The stopping condition for the above designed ANN is:

- Threshold value for error is 0.04.
- No of epochs are 4000.

Then it will be used to detect genres from emotions as input which will be unlabeled data. The neural network gives the value of each genre present in the trailer. The genres that can be detected are action, comedy, crime, horror, romantic, drama, and thriller. The maximum of all the genres detected from facial expressions of viewer for a movie is assigned as final genre for the viewer. Thus instead of manually assigning the genre to each movie, the system has tagged the movie with the genre evaluated from face expressions of all the viewers. The overall genre of the movie will be calculated based on the genre for which maximum value is attained from all the viewers. For example: if expressions of majority of the viewers indicate that the movie belongs to comedy genre then the movie will be tagged with this genre. ENCOG framework [7] for implementing machine learning techniques is used.

Of all the genres detected from facial expressions of viewer for a movie is assigned as final genre for the viewer (Table 2).

Table 2. Neural network architecture

Layer	Input neurons	Activation function
Input layer	8	Sigmoid function
Hidden layer	7	Sigmoid function
Output layer	7	Sigmoid function

where Sigmoid function (y) = $[1/1 + e^{-x}]$

Implementing neural network using ENCOG framework:

1. var SAMPLE= [anger, contempt, disgust, fear, happiness, neutral, sadness, surprise];
   //SAMPLE variable to store the testing input.

2. var network = ENCOG.BasicNetwork.create
   ([ENCOG.BasicLayer.create(ENCOG.ActivationSigmoid.create(), 8, 1),
   ENCOG.BasicLayer.create(ENCOG.ActivationSigmoid.create(), 7, 1),
   ENCOG.BasicLayer.create(ENCOG.ActivationSigmoid.create(), 7, 0)]);
   // "network" is defined here with the specified number of layers and neurons as well as activation function.

3. network.randomize();
   // Edge weights are randomly initialized.

4. var train = ENCOG.PropagationTrainer.create(network, INPUT, IDEAL,"RPROP", 0, 0);
   // "train" will store the properly trained artificial neural network with INPUT as the ideal inputs and OUTPUT as
   the target outputs used in supervised learning. RPROP (Resilient Backpropagation) algorithm is used for training

5. var iteration = 1;
   // iteration variable is initialized.

6. do
   {
   train.iteration();
   var str = "Training Iteration #" + iteration + ", Error: "+ train.error;
   iteration++;
   } while (iteration<4000 && train.error>0.04);
   // training is done with 4000 epochs and error threshold = 0.04 is used as stopping conditions.

7. var output = new Array(8);
   // "output" variable for storing testing outputs

8. network.compute(SAMPLE, output);
   //trained "network" is feed with the SAMPLE (test input) and test output is stored in "output".

9. var str = "Input: " + String(SAMPLE) + " Output: " + String(output);
   // for displaying test inputs and their corresponding outputs.

3.4   Sentiment Analysis of Review

Sentiment analysis of reviews are done using natural language processing, machine learning techniques, text analysis, statistical and linguistics knowledge to analyze, identify and extract information from documents [6, 15, 20]. There are two main approaches to sentiment analysis: the lexicon-based and the learning approach.

The lexicon-based approach calculates the semantic orientation of words in a text by obtaining word polarities from a lexicon [21]

Sentiment analyzer automatically extracts sentiments (positivity/negativity), opinion objects and emotions (liking, anger, disgust, etc.) from unstructured text information and provides the weight of each sentence in a review. The weight of each sentence is classified into positive weight, negative weight, and neutral weight also. After that, we are applying rating estimation algorithm for generating review rating based on review analysis. IMDB dataset [12] is used to provide reviews (positive, negative and neutral) on which the algorithm is tested and modified to incorporate the effect of all the reviews.

The review rating estimation algorithm is as follows:

Input: Review for a movie
Initialize all the parameters
1. Retrieve sentimental weights
2. Initialize variables
3. Find average for positive weights and negative weights.
4. Find max negative, positive and min negative, positive weight.
5. Scale average using min-max normalization for both positive and negative averages between 0 to 10.
6. Calculate balancing factor
  F factor = max sent count(pos, neg, neut)/min sent count(pos, neg, neut)
7. If (Pos sent count > Neg Sent count)
    Neg scaled avg = Neg scaled avg / F_{factor}
  else
    Pos scaled avg = Pos scaled avg / F_{factor}
8. Set initial neutral average to 5
9. Calculate neutral average using the formula :
Neut Avg = (ini neut Avg * (Neut Sent Count + Pos Sent Count - Neg Sent Count)) / Sum of all sentence count
10. Calculate rating by
If (Negative Sentence Count >2)
  Review rating =(Positive Scaled + Negative scaled)/2 + Neutral Average
else
  Review rating = (Positive Scaled + Negative scaled) + Neutral Average
Output: Review Rating for a review

The overall rating for a review can be calculated as:

Overall rating = (review rating + user rating)/2

The neutral review also affects the rating of a movie. Review for a movie can be like "The movie was not good but Salman's acting was amazing". Such a review gives neutral weight. To consider such reviews, neutral weight average is calculated. To improve the accuracy of the algorithm, rating from the user in stars has been explicitly taken. The overall rating by a user for a movie will be average of review rating and user rating. The final movie rating will be the mean average of the overall rating for all reviews of all users. This would then be used to recommend movies to other users.

Example: movie review [12] for film Haider [11]

I do not think we have any praise left for Tabu, KK, and Irrfan. They are marvelous. Now Haider! Well, I have seen Hamlet, as Kenneth Branagh comes pretty close to actual play. Its such a mammoth. I love Hamlet's soliloquy and that's what makes Hamlet good for me. Now Haider has that missing and whatever little was there was a bit disjointed. The initial part of a movie is pretty weak, Shahid's performance in the first half is very ordinary. Things change in second half, and Shahid has done a splendid job. The movie suddenly gets on with business and breathes again. I did not find Haider's descent into madness that convincing at all. I think VB still bit a more than he could chew. In past, he has done the brilliant and perfect job of taking these plays and making them even better and more desi. However, this time, it did not work that well. So I have rated movie at 6/10, although 6.5 but that was not possible on this IMDb scale. Btw best thing about the movie was Bismil song and Aao Na, I wish there were more of these [11].

User rating from the review is computed as Fig. 5. and the user provided 6 stars. Therefore,

$$\text{Rating} = (\text{review rating} + \text{user rating})/2$$
$$= (6.7 + 6)/2 = 6.35$$

4   Experimental Results

Below are the screenshots of result obtained when a user tries to access our Website.

In Fig. 6(a) registering a new user is a 2-step process. In the first step registration form will be displayed on the screen in which user needs to enter his credentials like username, email-Id, password. In the second step, the image of registering user will be captured and stored in the database so that it can be used to verify a user at the time of login. From the image, age and gender of registering a user are evaluated.

In Fig. 6(b) the registered user will login to his/her account using the user-defined username and password then on clicking "verify yourself" the user will be prompted to capture his/her image. For verification, the image will be compared to the registered image stored in our database. If all the credentials and the image are verified, then the user can successfully access the account and is directed to the main site.

In Fig. 7(a) the user's captured image will be compared with the user's original image used while creating the account. If the user is verified, he will be redirected to his/her account.

```
Sentences with sentiment objects and phrases:

Sentence Weight = 1
Sentence Weight = 1
Sentence Weight = 0
Sentence Weight = 1
Sentence Weight = 0
Sentence Weight = 1.3692307692308
Sentence Weight = -2.9812270398519
Sentence Weight = -2.8168113585237
Sentence Weight = 1
Sentence Weight = 0
Sentence Weight = -2
Sentence Weight = -2
Sentence Weight = 1.6391769149382
Sentence Weight = 0.048737017882858
Sentence Weight = 0
Sentence Weight = 1
Sentence Weight = 0
Positive Weight Average = 1.0071430877565
Positive Sentence Count= 8
Positive High= 1.6391769149382
Positive Low= 0.048737017882858
Negative Weight Average = -2.4105095995939
Negative Sentence Count= 4
Negative High= -2
Negative Low= -2.9812270398519
Neutral Weight = 5
Neutral Sentence Count= 5
Neutral Scaled Average = 5
Positive Scaled Average = 6.0260439369514
Negative Scaled Average = 5.418903257479
Factor = 2
Factor2 = 0.47058823529412
Neutral Scaled Average = 2.3529411764706
Positive Scaled Average = 6.0260439369514
Negative Scaled Average = 2.7094516287395
Rating = 6.7206889593161

Categorized Opinions with sentiment polarity (positive/negative)
```

Fig. 5. Sentiment analysis of review for movie Haider

In Fig. 7(b) the captured image age and gender of the user is detected by analyzing the facial features. The system will recommend movies based on movie preferences of users in database belonging to the same age group and gender. The movie list is sorted by movie rating evaluated from reviews taken by the user. The movies are also recommended by genres that the user prefers to watch on our website.

In Fig. 8 the navigation menu on the left side of the screen provides all the available options to a user to interact with the system. Each section like Latest Movies, Watch History displays a list of movies to a user.

In Fig. 9 the Latest section, the system recommends the latest movies to the user. The movie list is arranged by year of release and average rating. This eliminates the problem of cold start which arises due to insufficient rating for a movie that has

(a)   (b)

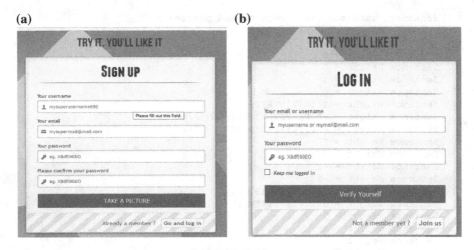

Fig. 6. (a) Registering the user (b) Login process

(a)   (b)

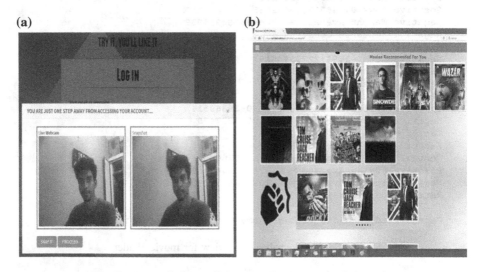

Fig. 7. (a) Login by verifying face (b) Recommended movies

released recently and has not gain much popularity amongst users till now but may prove to be a movie worth watching.

In Fig. 10(a) the Watch History section enables user to view all the movies he has watched in past on our website.

In Fig. 10(b) the Trending Section displays the popular movies i.e. the movies that are watched by the maximum users on the website. The list is sorted in descending order by total number of views.

In Fig. 11(a) whenever a user watches a movie, all the related information about movies and user goes into the database. From the database, all the genres that user

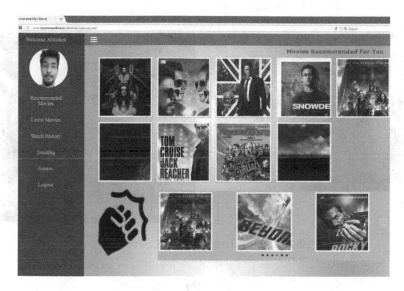

Fig. 8. Navigation menu

Fig. 9. Latest Movies

frequently watches like comedy, action can be calculated from her history. Thus the "Recommended For You" section suggest the movies belonging to all the genres that a user is interested in. For Example: Suppose a user 'Deepika' frequently watches comedy and romantic movies on our website, so she will be recommended movies which are sorted by average rating in these categories.

(a)   **(b)**

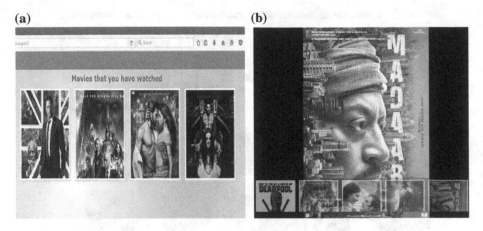

Fig. 10. (a) Watch History (b) Top viewed movies

(a)   **(b)**

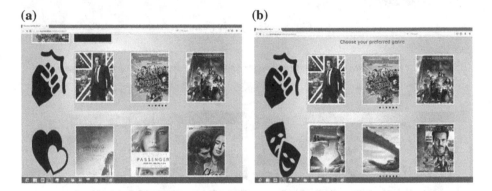

Fig. 11. (a) Preferred genres (b) Genre section

In Fig. 11(b) at times a user might be interested in trying something new. For example, if a user usually watches comedy movies, she may wish to watch a movie of drama genre. To satisfy such needs, a genre section is created where movies belonging to all genres other than what user frequently watches are recommended.

In Fig. 12 Once the user selects a poster of a movie he would like to watch, he\she will be directed to watch the trailer of the movie. As soon as he starts the trailer, the webcam gets on and starts recording his facial expressions at regular intervals of time to tag the movie with the genre. The genre is evaluated by recorded emotions.

In Fig. 13 contains the graph and a pie chart that depicts the emotions associated with facial expression of the user thus records the user's emotional reaction of the user while watching the trailer. The pie chart shows the percentage of each of the eight emotions: Happiness, sadness, neutral, anger, surprise, contempt, disgust, fear.

In Fig. 14 displays the value of each of the seven genres associated with the trailer. The genres are Horror, Romance, Drama, Crime, Comedy, Action, and Thriller. The genres with the maximum area in the pie chart are the probable genre of the movie.

Fig. 12. Trailer of selected movie

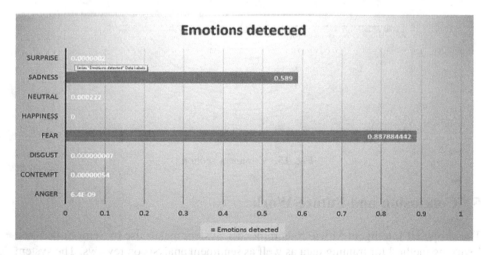

Fig. 13. Emotions detected

In Fig. 15 the system asks the user to give review and rating for the movie based on his experience once he has watched the trailer. From the review and star rating, user rating is calculated. The overall rating for a movie is the average of all user ratings.

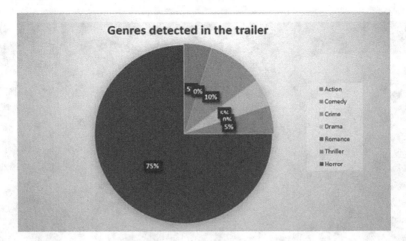

Fig. 14. Summarization of genres

Fig. 15. Comment section

5 Conclusion and Future Work

The proposed Intelligent Movie Recommender System makes use of Semi-Supervised Learning method for training data as well as sentiment analysis on reviews. The system facilitates a web-based user interface i.e. a website that has a user database and has a Learning model tailored to each user. This interface is dynamic and updates regularly. Afterward, it tags a movie with genres to which they belong based on expressions of users watching the trailer. The major problem arises with this technique is when the viewer gives neutral face expressions while watching a movie. In this case the system is unable to determine the genre of the movie accurately. The recommendations are refined with the help of reviews and rating taken by the users who have watched that movie.

A user is allowed to create a single account, and only he can log in from his account as we verify face every time. The accuracy of the proposed recommendation system can be improved by adding more analysis factor to user behavior. Location or mood of the user, special occasions in the year like festivals can also be taken into consideration to recommend movies. In further updates text summarization on reviews can be implemented which summaries user comment into single line will comments. Review Authenticity can be applied to the system to prevent fake and misguiding reviews. Only genuine reviews would be considered for evaluation of movie rating. In future, the system can be used with nearby cinema halls to book movie tickets online through our website [22]. Our approach can be extended to various application domains to recommend music, books, etc.

References

1. Roy, D., Kundu, A.: Design of movie recommendation system by means of collaborative filtering. Int. J. Emerg. Technol. and Adv. Eng. **3** (2013) http://www.ijetae.com/files/Volume3Issue4/IJETAE-0413-12.pdf
2. Ho, A.T., Menezes, I.L.L., Tagmouti, Y.: E-mrs: emotion based movie recommender system. In: Proceedings of IADIS eCommerce Conference. University of Washington Both-ell, USA, pp. 1–8 (2006)
3. ANN: Definition of artificial neural networks (2013). http://briandolhansky.com/blog/artificial-neural-networks-linear regression-part-1
4. API: Microsoft API for emotion detection (2015a). https://www.microsoft.com/cognitive-services/en-us/emotion-api/documentation
5. API: Microsoft API for face analysis (2015b). https://dev.projectoxford.ai/docs/services/563879b61984550e40cbbe8d/operations/563879b61984550f30395236
6. Pang, B., Lee, L., Vaithyanathan, S.: sentiment classification using machine learning techniques. Comput. Intell. **22**, 110–125 (2006)
7. ENCOG: ENCOG framework for training data-set (2016). http://www.heatonresearch.com/encog/
8. ENCOG: Encog framework for training data-set (2016)
9. Arora, G., Kumar, A., Devre, G.S., Ghumare, A.: Movie recommendation system based on user's similarity. IJCSMC J. **3**, 765–770 (2014). Springer
10. Adomavicius, G., Tuzhilin, A.: Context-aware recommender systems. In: Ricci, F., Rokach, L., Shapira, B., Kantor, P.B. (eds.) Recommender Systems Handbook, pp. 217–253. Springer, New York (2010). http://link.springer.com/chapter/10.10072F978-0-387-85820-3_7
11. HAIDER: review from haider movie (2014). http://www.imdb.com/title/tt3390572/reviews?ref_=tt_ql_3
12. IMDB: Data-set for sentiment and review analysis (1990). http://www.imdb.com
13. Wakil, K., Ali, K., Bakhtyar, R., Alaadin, K.: Improving web movie recommender system based on emotions. (IJACSA) Int. J. Adv. Comput. Sci. Appl. **6** (2015)
14. Madadipouya, K.: A location-based movie recommender system using collaborative filtering. Int. J. Found. Comput. Sci. Technol. (IJFCST) **5** (2015)
15. Kennedy, A., Inkpen, D.: Sentiment classification of movie reviews using contextual valence shifters. Int. J. Comput. Appl. **22**, 1467–8640 (2006)
16. Kumar, M., Yadav, D.K., Singh A., Gupta, V.K.: A movie recommender system: Movrec. In: International Journal Of Computer Applications, ACL 2002 Conference on Empirical Methods in Natural Language Processing, vol. 124, pp. 0975– 8887 (2015)

17. Amini, M., Nasiri, M., Afzali, M.: Sentiment classification using machine learning techniques. (IJCSIS) Int. J. Comput. Sci. Inf. Secur. 12 (2014)
18. Samples, Data-set for making neural network (2016). www.recommendme.in/abhishek/neuralnetwork
19. Science Direct. Definition of cold start problem (2016). http://www.sciencedirect.com/science/article/pii/S0306437914001525
20. Kim, S.-M., Hovy, E.: Determining the sentiment of opinions. In: Proceedings of the COLING conference (2004). http://www.isi.edu/natural-language/people/hovy/papers/04Coling-opinionvalences.pdf
21. Sentiment analysis API. In: Natural Language Processing, vol. 124, pp. 0975–8887. http://www.intellexer.com/sentiment_analyzer.html/methods
22. Link of the website that we have created: http://www.recommendme.in/abhishek/startpage/

A Comparative Study of Classifier Performance on Spatial and Temporal Features of Handwritten Behavioural Data

Asok Bandyopadhyay[✉] and Abhisek Hazra

Centre for Development of Advanced Computing-Kolkata, Plot-E2/1, Block-GP,
Sector-V, Salt Lake City, Kolkata 700091, West Bengal, India
asok.bandyopadhyay@cdac.in

Abstract. The issue of comparing classification algorithms on a data set to find optimal classifier has always been a demanding issue in Machine Learning studies. The goal of this work was to compare the performance of different classifiers tested on spatio-temporal features used to distinguish between true and distorted handwriting behaviour to detect deception using machine learning experiments. Dynamic handwriting features depict a writer's individuality and neuro-psychic condition. Handwritten features share inter-class characteristics while a few of them are very unique to an individual writer. However, extraction of discriminating features & selecting them in decision making process with respect to an individual is really crucial and complex thing, which helps the construction of any classifier, aiming to build the decision support system. Spatio-temporal attributes carry significant discriminatory information in deception detection. In this work, a set of total twelve features (spatial and temporal with pressure measures) features were used from online handwritten data. Five different classifiers (Naïve Bayes, Logistic Regression, Multi-Layer Perceptron, Random Forest and Support Vector Machine) were tested on these feature set with three separate factual descriptions: Person, Action & Event, in both true and distorted mode to develop a decision support system based on the outcome of the experiment on handwriting behaviour. Both Support Vector Machine and Logistic Regression technique have shown good performance in distinguishing between true and distorted handwritten contents.

Keywords: Handwriting behaviour · Machine learning · Spatio-temporal feature · Classifier

1 Introduction

The main aim of our activity was to develop a perception based automated decision support system for handwriting behaviour analysis applicable in the area of deception detection. We have tried to evaluate brain-hand performance, as manifested through handwriting behaviour which is a valid measure for detecting the dis-automaticity that is indicative of deception detection in forensic field. The cognitive approaches explaining detection assume that encoding a deceptive message requires a greater cognitive effort than telling the truth because of higher processing capacity demands particularly when the lie involves a report about a complex event.

© Springer International Publishing AG 2017
A. Basu et al. (Eds.): IHCI 2016, LNCS 10127, pp. 111–121, 2017.
DOI: 10.1007/978-3-319-52503-7_9

In case of Lie detection, suspected person may be identified by asking him to write a statement about a particular incident in which his involvement is suspected. A typical pattern will emerge through the analysis of perceptual behavioural pattern of the suspect which will reflect the stressed condition of the person and it may be used as an aid to detect whether a person is lying or not [1].

The computerized system makes it possible to compare handwriting under different conditions; therefore we have compared the handwriting of the same individuals when asked to write truthful and deceptive sentences. Our research hypothesis is that differences will be found between writing of truthful sentences and writing of false sentences in pressure, temporal (stroke duration on digitizer and in air) and spatial measures (strokes path length, height and width) obtained by the computerized system [2]. Based on the finding of the above clinical studies we can predict that in deceptive writing, the mean and standard deviations of handwriting measures of each participant will be varied. Thus while writing deceptive sentences, higher pressure will be implemented, longer duration time per stroke (on paper and in air) will be required, and letter strokes will be larger in comparison to truthful writing.

1.1 Objective

The aim of the experimental research is multi folded. The most important part of this activity was to study the sensory, motor and cognitive factors affecting the handwriting of a person along with creation of sample database for different person's handwriting. Next part of the task was to develop the perceptual experimental setup for analyzing handwriting behaviour. Finally, the target is to utilize the research outcome by developing a prototype of decision support system for handwriting behaviour analysis to detect deception. In doing so, we needed to select an optimal classifier to work on the spatio-temporal feature sets [3]; the work has been executed and reported here.

2 Data Collection Methodology

The handwritten data collection procedure was carried out in two phases. The strategies were discussed with domain experts before commencing the field data collection camp. Initially, to collect offline data, the template and guidelines for offline hand-writing data was developed prior to the data collection camp. The contributors were guided throughout the process for seamless data collection. They were asked to narrate the most memorable incident of their life in true modality as well as to deliberately distort the same event by writing distorted sentences. The aim of this work was to find any subtle changes between true and distorted behaviour throughout the handwriting images by computing morphological and shape based image features.

Altogether we have collected online and offline handwritten data set from 148 students in two data collection camps. A video with proper subtitle was shown to all the students. This 8-minutes video was shown only once. After completion of the video session, we proceeded to the next phase of collecting true mode offline data. The offline template contained the three basic parametric information of a video namely, persons,

Persons along with their actions and events. The template for offline data collection contained separate bullet points for the above mentioned categories because distortions were easily captured in that process. Next task was to collect true mode online data using Wacom's Intuos Pro tablet. During this process, the previous offline handwritten statements were not shown to the students. Afterwards, the distorted mode online data was collected. Students were asked to describe three parameters (persons, Persons along with their actions and events). Finally, the distorted mode offline data was captured. Carbon papers were attached with the offline data collection template to get the underlined statements which were distorted by the students.

Front-end of the automated handwriting behaviour analysis tool is shown below (Fig. 1).

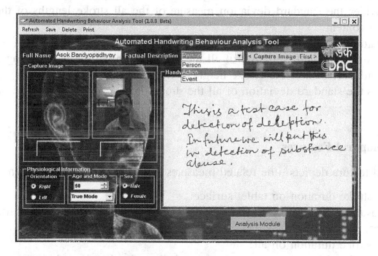

Fig. 1. Interafce for automated handwriting behaviour analysis tool

Online data collection were performed using a WACOM Intuos pro digitizing tablet, a stylus with a pressure-sensitive tip having pressure sensitivity of 0–2048 levels and a sampling rate of 200 points per second [4]. The images of the data contributors were captured using a Logitech HD webcam, although facility for capturing image using any USB camera is provided in our data collection software.

3   Pre-processing and Feature Extraction

Processing time series handwriting data before feature extraction is an important step which was accomplished in our data collection software. The steps involved are duplicate point elimination; noise removal caused by hand trembling etc. Feature extraction module consists of spatial and temporal information along with stylus pressure measures as obtained from dynamic handwriting data. The online handwriting analysis consists of twelve such spatial and temporal features. Next section describes them in details.

3.1 Spatial Features

Spatial features are related to distance measure.

a. Mean stroke length
   It measures the mean length of each stroke from its beginning point to end point in a single writing panel.
b. Mean stroke width
   It measures the mean of the width of each stroke beginning from start to end point.
c. Mean stroke height
   It is obtained by measuring the mean of the heights of each stroke from start to end point.
d. Standard deviation of the stroke lengths
   It provides the standard deviation measure of the all stroke lengths of the handwriting content.
e. Standard deviation of the stroke width
   It gives the standard deviation of all the stroke widths of the handwriting content.
f. Standard deviation of the stroke heights
   It gives the standard deviation of all the stroke heights of the handwriting content.

3.2 Temporal Features

Temporal feature depicts time related measures from the handwritten content.

a. Mean stroke duration on tablet surface:
   It measures the mean of all stroke durations during pen down time on the tablet surface.
b. Mean stroke duration on air:
   It is calculated by measuring the mean of all stroke durations during pen up time over the tablet surface.
c. Standard deviation of the stroke durations on tablet surface:
   It gives the standard deviation of all stroke durations during pen down time on the tablet surface.
d. Standard deviation of the stroke duration in air:
   It gives the standard deviation of all stroke durations during pen up time over the tablet.

3.3 Pressure Measure

The absolute pressure measure exerted by the stylus on tablet surface is captured and utilized. Each stroke contains some points (X, Y coordinates and associated pressure with each point) with timestamp. During the pen down time, a pressure value is obtained for each stroke points, while during pen up time this value becomes zero as there is no contact between stylus and tablet. It is has been found that in spite of inter-class and intra-class variabilities among writers, a significant difference in pressure level can be found in case of true and distorted behaviour of the writers [5, 6].

Figure 2 represents descriptive statistical information obtained using IBM SPSS version 23.0 from handwritten data sets of field data collection camp. A total set of 114 student's data (N = 228 taking true & distorted mode) was collected. The feature names are described in the first column while second column describes the total number of data samples. 'Minimum' and 'Maximum' denote the minimum and maximum value of that specific feature. 'Mean' & 'Std. deviation' represent the arithmetic mean and standard deviation of that particular feature vector.

Descriptive Statistics

	N	Minimum	Maximum	Mean	Std. Deviation
Mean Pressure	228	69.63901450	402.7030470	229.4150852	69.67044137
Standard Deviation of Pressure	228	97.37527364	356.6785956	218.4369326	53.15881654
Mean Stroke Duration on Tablet Surface	228	484.3529412	12077.99408	2994.058601	1670.679315
Mean Stroke Duration In Air	228	.467143787	10.37322454	1.692027995	1.113627177
Standard Deviation of the Stroke Durations on Tablet Surface	228	86414.81661	43470308.59	3629439.240	4656204.310
Standard Deviation of the Stroke Duration in Air	228	47.00884240	72428.66487	2464.704403	6369.715042
Mean Stroke Length	228	22.10609268	515.7648894	154.3993623	72.69066844
Mean Stroke Width	228	5.014285714	35.42857143	13.93707280	5.603814347
Mean Stroke Height	228	0.590153040	38.71428571	17.56651908	4.344402995
Standard Deviation of the Stroke Lengths	228	433.2584814	264190.8191	28626.64834	32244.18470
Standard Deviation of the Stroke Widths	228	6.748888889	1095.862400	218.7170726	204.0120656
Standard Deviation of the Stroke Heights	228	32.03242530	574.1997622	165.3749458	94.90354591
Valid N (listwise)	228				

Fig. 2. Statistical description of online handwriting behavioural data set

4   Performance Evaluation

The statistical significance of a classifier is compared to other classifiers on handwritten behavioural data. For each classifier, training and testing sample randomly drawn from the base set using 10-fold cross validation technique [7], the model was trained, and tested. We have segmented our data set based on three factual descriptions; person (true and distort), action (true and distorted) and event (true and distorted). Five different classifiers (Simple Naïve Bayes, Logistic Regression, Multi-Layer Perceptron, Random Forest and Support Vector Machine) were tested on our dataset for different factual descriptions for a comparative output.

The above diagram (Fig. 3) reveals that simple logistic regression based classifier give better performance among the classifiers. In case of action factual description (Fig. 4), Support vector Machine (SVM) gives better performance compared with other classifiers. In the case of event factual description, it was observed that mean accuracy

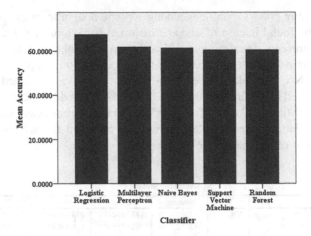

Fig. 3. Comparison of classifiers for person factual description (true and distorted)

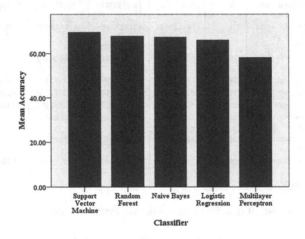

Fig. 4. Comparison of classifiers for action factual description (true and distorted)

measure for simple logistic regression based classifier performs better than other classifiers we have chosen, though SVM's mean accuracy is quite closer in this case (Fig. 5).

Further, we have segregated true and distorted modalities of data based on three factual descriptions person, action & event independently. The result of classification in terms of True positive (TP), False Positive (FP), Precision Recall and Fscore of person (true), person (distorted), action (true), action (distorted), event (true) & event (distorted) is reported below.

The following Fig. 6 depicts the comprehensive results obtained from this experimentation. It is observed that in person factual description modality, the true positive measures are 0.667 and 0.684 for simple Logistic Regression classifier in true & distorted mode respectively. In distorted mode, the true positive rate for SVM is 0.684,

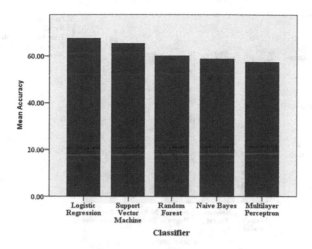

Fig. 5. Comparison of classifiers for event factual description (truc and distorted)

	Naïve Bayes		Logistic Regression		Multi-layer Perceptron		Random Forest		Support Vector Machine		
	TRUE	DISTORTED	TRUE	DISTORTED	TRUE	DISTORTED	TRUE	DISTORTED	TRUE	DISTORTED	
TP Rate	0.561	0.667	0.667	0.684	0.614	0.623	0.623	0.588	0.526	0.684	
FP Rate	0.333	0.439	0.316	0.333	0.277	0.386	0.412	0.377	0.316	0.474	
Precision	0.627	0.603	0.679	0.672	0.619	0.617	0.602	0.609	0.625	0.591	
Recall	0.561	0.667	0.667	0.684	0.614	0.623	0.623	0.588	0.526	0.684	
Fscore	0.593	0.633	0.673	0.678	0.617	0.62	0.612	0.598	0.571	0.634	**Person**
TP Rate	0.655	0.693	0.646	0.675	0.531	0.632	0.673	0.684	0.566	0.825	
FP Rate	0.307	0.345	0.325	0.354	0.368	0.469	0.316	0.327	0.175	0.434	
Precision	0.679	0.669	0.664	0.658	0.588	0.576	0.679	0.678	0.762	0.657	
Recall	0.655	0.693	0.646	0.675	0.531	0.632	0.673	0.684	0.566	0.825	
Fscore	0.667	0.681	0.655	0.667	0.558	0.603	0.676	0.681	0.65	0.732	**Action**
TP Rate	0.36	0.816	0.64	0.711	0.605	0.544	0.623	0.579	0.535	0.772	
FP Rate	0.184	0.64	0.289	0.36	0.456	0.395	0.421	0.377	0.228	0.465	
Precision	0.661	0.56	0.689	0.664	0.57	0.579	0.597	0.606	0.701	0.624	
Recall	0.36	0.816	0.64	0.711	0.605	0.544	0.623	0.579	0.535	0.772	
Fscore	0.466	0.664	0.664	0.686	0.587	0.561	0.609	0.592	0.607	0.69	**Event**

Fig. 6. Comparison of classifiers for person, action & event factual description

which matches with Logistic Regression. In action modality, Random Forest classifier has the highest TP rate of 0.673 in true mode while SVM has the highest TP rate of 0.825 in distorted mode. Finally, Logistic Regression classifier has the highest TP rate of 0.64 in true mode while Naïve Bayes classifier has the highest TP rate of 0.816 in distorted mode for event factual description. The diagrams below (Figs. 7, 8, 9, 10, 11 and 12) shows the comparison bar charts of all the classifiers for their accuracy measures by plotting precision, recall, TP rate, FP rate and FScore values of person true, person distorted, action true, action distorted, event true & event distorted handwriting data [8].

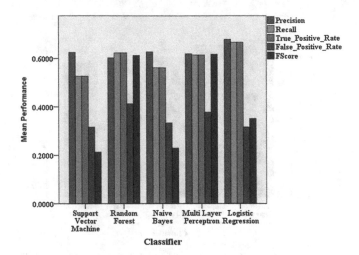

Fig. 7. Comparison of classifiers for person true factual description

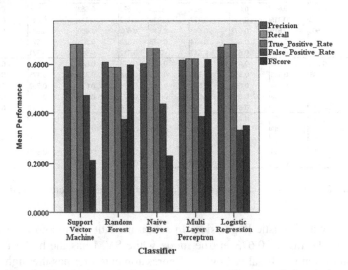

Fig. 8. Comparison of classifiers for person false factual description

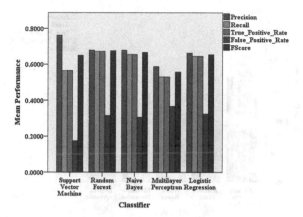

Fig. 9. Comparison of classifiers for action true factual description

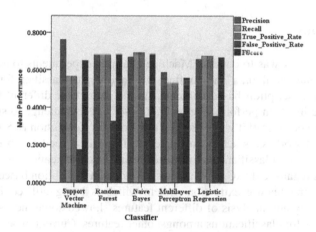

Fig. 10. Comparison of classifiers for action false factual description

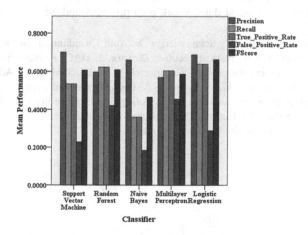

Fig. 11. Comparison of classifiers for event true factual description

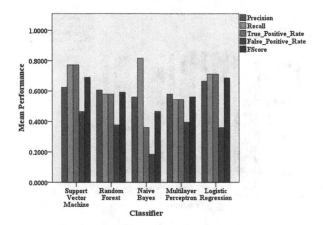

Fig. 12. Comparison of classifiers for event false factual description

5   Conclusion

The aim of this work was to conduct Machine Learning experiments to analyze online handwriting behavioural data and to find out an optimal classifier for segregating different factual description based contents. As the dataset for different factual mode show difference in Mean performance level, it poses a real challenge to select a single classifier. However, overall it was observed that Logistic Regression & Support Vector Machine (SVM) work satisfactory in our case. There is a need to observe whether combining these two classifiers yields better result in classification or not. Work is going on in this regard. C-DAC, Kolkata has developed a Perception based handwritten data collection and feature extraction system which is being utilized for field data collection. A study and analysis of different features showed some merits in spatial & temporal features for classifications amongst other features. Current experimental study will help to choose the suitable classifier for developing the decision making system. However, it is necessary to perform the same analysis on derived and local features of online handwritten behavioural data. Work is in progress and will be reported shortly.

Acknowledgements. We sincerely acknowledge Dr. Amit Chaudhuri, Joint Director, C-DAC, Kolkata and Col. (Retd.) A.K. Nath, Executive Director, C-DAC, Kolkata for their constant support and help in executing the project. We sincerely thank Dr. Subhranshu Aditya, School of Cognitive Science, JU for his valuable inputs in designing perceptual experimental set up. We also thank CID Guwahati, Assam and Don Bosco College and Engineering, Guwahati, Assam for providing support for Data Collection. The project is funded by DeitY (Department of Electronics and Information Technology), Govt. of India.

References

1. Osborn, A.S.: Questioned Documents, 2nd edn., pp. 205–216, 226–233, 247–248, 363–376. Nelson-Hall, Chicago (1929)
2. Rosenblum, S., Parush, S., Weiss, P.L.: The in air phenomenon: temporal and spatial correlates of the handwriting process. Percept. Mot. Skills **96**(3), 933–954 (2003)
3. Rosenblum, S., Parush, S., Weiss, P.L.: Computerized temporal handwriting characteristics of proficient and non-proficient handwriters. Am. J. Occup. Ther. **57**(2), 129–138 (2003)
4. http://www.wacom.com/en-gb/products/pen-tablets/intuos-pro-medium#Specifications
5. Mavrogiorgou, P., Mergl, R., Tigges, P., El Husseini, J., Schroter, A., Juckel, G., Zaudig, M., Hegerl, M.: Kinematic analysis of handwriting movements in patients with obsessive-compulsive disorder. J. Neurol. Neurosurg. Psychiatry **70**(5), 605–612 (2001)
6. Srihari, S.N., Cha, S.-H., Arora, H., Lee, S.: Individuality of handwriting. J. Forensic Sci. **47**, 856–872 (2002)
7. Alpaydin, E.: Combined 5×2 CV F test for comparing supervised classification learning algorithms. Neural Comput. **11**, 1885–1892 (1999)
8. Demsar, J.: Statistical comparison of classifiers over multiple data sets. J. Mach. Learn. Res. **7**, 1–30 (2006)

Estimation of Mental Fatigue During EEG Based Motor Imagery

Upasana Talukdar[✉] and Shyamanta M. Hazarika

Biomimetic and Cognitive Robotics Lab, Tezpur University, Tezpur, India
upasanat123@gmail.com, smh@tezu.ernet.in

Abstract. Mental Fatigue is a cognitive state which is an outcome of labour or protracted exercise finally leading to downgrading of mental performance. This leads to reduction in efficiency and disinclination of motor skills. Analysis of mental fatigue thus becomes momentous for assessing one's capability. The aim here is to analyse whether motor imagery is fatiguing. There are many parameters to evaluate fatigue. This study analyses parietal alpha and frontal theta in motor imagery tasks. Decrement of arousal level, working memory and information encoding have been proven to be associated with increased theta power. Increase in alpha power indicates increase in mental effort to maintain vigilance level. When a person experiences fatigue, their concentration, attention, focus and vigilance level decreases for which they need to put more attention which leads to increase in alpha power. We exploit these EEG oscillatory rhythm fluctuations to model EEG-fatigue relationships. A statistical classifier is used to model EEG-fatigue relationship accurately. With Kernel Partial Least Square output we track the growth of mental fatigue with time.

Keywords: Mental fatigue · EEG · Kernel partial least squares · Motor imagery

1 Introduction

Mental fatigue is a general feeling of tiredness, reduced efficiency and alertness, feelings of inhibition and impaired mental performance [4]. Trezo et al. [1] defines mental fatigue as subjective feelings of low energy and motivation, accompanied by weakness. However, there is no particular definition of fatigue. The definition of mental fatigue varies according to the area where it has been defined. In general, fatigue can be described as a sensation of exhaustion, weariness, loss of energy, weakness or low arousal level.

The main effect of mental fatigue are slow cognition, negligence in activities, declination of performance, reduction in focus, difficulty in memorizing and learning new things [5]. Analysis and estimation of mental fatigue thus becomes substantial to assess one's cognitive competence. Literature reports different techniques to estimate mental fatigue. Till date, the most convenient and easiest way to estimate mental fatigue is the self report ratings based on questionnaires. Different scales are being used to analyze and evaluate mental fatigue. In addition to this, observed behavior while performing a particular task, response time and response accuracy can also be used to

© Springer International Publishing AG 2017
A. Basu et al. (Eds.): IHCI 2016, LNCS 10127, pp. 122–132, 2017.
DOI: 10.1007/978-3-319-52503-7_10

estimate the growth of mental fatigue. *Response time* is the time period taken by a participant to react to a given stimulus/event. *Response accuracy* is the degree of closeness of measurement of a response to that response's correct value. However, these measures also suffer from limitations. Subjective self report measures are not a tenacious mode of measurement as subjects may not be able to report accurately what they feel. Likewise observed behavior may not be able to capture the degree of fatigue level of a participant. The response time and accuracy may not be able to estimate fatigue accurately as an inefficient person may take a long time to respond at the very beginning of a task. Recent studies have shown that Electroencephalography (EEG) signals can be used to analyze mental fatigue [1, 4, 6, 7, 9, 10, 14, 18]. Several features of EEG have found its place in estimation of mental fatigue.

The paper aims to check whether a person experiences fatigue while performing motor imagery (MI) tasks i.e. whether motor imagery is fatiguing. In this study, we isolate the low and high fatigue states in motor imagery data. The study uses two EEG band powers as identified in the literature: *parietal alpha* and *frontal theta*, to estimate mental fatigue. Increased theta power indicates decrement of arousal level, working memory and information encoding while alpha power increases with increase in effortful attention which participants put to maintain the vigilance level while experiencing fatigue. Hence, the idea is to track the increase in alpha and theta power with time enabling one to track the growth of fatigue.

To estimate mental fatigue, Kernel partial least square (KPLS) is used. The idea is closely derived from [1]. KPLS gives the regression coefficients between the class labels - low fatigue and high fatigue and the frequency band power - theta and alpha power; thus portraying the relevance between the two. This shows the relationship of the two band powers with the two class labels; portraying the changes in these two band powers in case of motor imagery data.

2 Related Work

2.1 Correlation of EEG and Different Cognitive States

EEG based estimation of cognitive states has attained much interest [1, 4, 6–12, 14, 16–20]. Literature identifies several observations based on EEG and power spectral analysis for estimating different cognitive states. Band power features have find its place in estimating different cognitive states like fluctuation in focus level, attention level, workload, fatigue, sleepiness etc. To study brain activities in context of cognitive states, literature discusses 4 primary locations of brain, i.e. temporal, frontal, occipital and parietal regions [11, 12]. EEG electrodes placed on the scalp captures the fluctuated voltages. For different level of brain activities, the neurological activity in the brain cause different frequencies [12]. As for instance, brain radiates lower frequency signals in asleep condition and higher frequency in awake condition [12, 13]. Thus there exists distinct correlation between frequency band power extracts and different cognitive states [12]. Table 1 portrays these relations as identified in the literature.

Table 1. Correlation of band power extracts and different cognitive states as identified in the literature

Frequency bands	Associated cognitive states
Delta (0.1–3.9 Hz)	Deep sleep is related with increase in delta power. More prominent at temporal lobe [12]
Theta (4–7.9 Hz)	Theta wave is associated to work memory, sleep and cognitive performance. Increased theta activity is related with decrement of performance, including information encoding, working memory and so forth. The band power increases with increase in fatigue [1, 9]
Alpha (8–12.9 Hz)	Alpha waves appears in the relaxed and effortless alertness. The increased alpha activity is related with an increase in mental effort. The band power increases with increase in fatigue [9]. More prominent at parietal lobe [1]
Beta (13–29.9 Hz)	Increased beta activity is related with increasing alertness, arousal and excitement [9]. More pronounced in temporal and occipital lobe [12]
Gamma (30–100 Hz)	It is associated with hypertension, high mental activity and indicative of information processing [12]. More pronounced at Centro- midline of the brain [12]

2.2 EEG and Mental Fatigue

Mental fatigue is a cognitive state which is related to other different cognitive states like attention level, workload level, focus level, sleepiness and so forth. This is because when a person experiences fatigue, his attention and focus level decreases, sleepiness and drowsiness increases, workload increases accompanied with downgrading of cognitive performance. Recent studies report different measures to estimate mental fatigue. The EEG band power extracts play a substantial role in analyzing and estimating mental fatigue. Borghini et al. [4] portrayed different aspects of EEG signals to estimate "mental workload", "mental fatigue" and "situational awareness". Borghini et al. summarize the main neuro-physiological findings in context of measurements of driver's brain activities during drive performance and how these can be correlated with the above mentioned three concepts; while Holm et al. [8] showed the way to estimate "mental workload" and "mental fatigue" using EEG. Different parameters for estimation of mental fatigue using EEG while performing motor imagery and emotional stimulation task have also been put forwarded by Pomer-Escher et al. [6]. They have depicted 14 different aspects of brain activity like θ, α, β, their ratios, ratios of their sums etc. that can be used to estimate mental fatigue. Cao et al. [9] investigated the changes in Steady state visually evoked potential (SSVEP) properties, EEG indices like amplitudes in β, α, θ and δ frequency bands and the amplitudes of their ratio indices θ/α and $(\theta + \alpha)/\beta$ with the rise in fatigue level. It found that β, α, θ, δ and $(\theta + \alpha)/\beta$ power increases while θ/α power decreases with increasing fatigue level. Jap et al. [10] assessed the four electroencephalography (EEG) activities, δ, θ, α and β along with α/β, θ/β, $(\theta + \alpha)/\beta$ and $(\theta + \alpha)/(\alpha + \beta)$ as indicators for fatigue detection. The paper shows how all these parameters show pronounced changes with increase in fatigue level. δ, θ, α and β along with α/β and $(\theta + \alpha)/\beta$ increases while θ/β and $(\theta + \alpha)/(\alpha + \beta)$ decreases as the person experiences more fatigue. Trezo et al. [1] showed how frontal theta and parietal alpha can be used to track the development of mental fatigue using EEG. They have used Kernel Partial Least Squares (KPLS) Algorithm to model EEG-fatigue relationship. It is found that with increase in fatigue, beta power was found to rise in 2 studies, decline in 2 and no change in 2 studies. Alpha power (investigated in seventeen studies) was found to rise in fifteen studies and drop in 2 studies. Theta activity out of sixteen studies was found to rise significantly in fourteen studies and no change in two studies. Delta power was found to increase in 4 studies, decline in 1 study and no change in one study [14].

Based on the related work done so far, this paper uses two EEG oscillatory rhythms: theta and alpha to model EEG-fatigue relationship.

3 Materials and Methods

3.1 Dataset Used

Motor imagery data from BCI Competition IV Dataset 2a[1] is used. EEG data available is recorded using twenty two EEG channels and three EOG channels from nine subjects.

3.2 Protocol of the Dataset

The subjects were asked to perform 4 different motor imagery tasks: right hand movement, left hand movement, tongue movement and both feet movement. For each subject EEG data was recorded in two sessions on two different days. Each session comprises of six runs. Runs are separated by short breaks. One run consists of 48 trials each of 6 s in length giving a total of 288 trials per session. Figure 1 shows the timing diagram of one such session. The total duration of the experiment is approximately 45 min. in length.

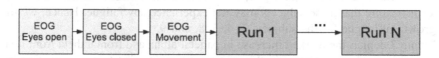

Fig. 1. Timing diagram of one session

Each trial begins with (at t = 0 s) a fixation cross appearing on the screen which tells the subjects to get ready for the task. It is followed by the appearance of a cue in the form of an arrow that asks the participants to accomplish the desired motor imagery task i.e. left arrow prompts subjects to imagine left hand movement, right arrow indicates right hand movement, up and down arrow for foot and tongue movement respectively. The subjects then accomplish the task till the fixation cross disappears at t = 6 s. Figure 2 depicts the timing scheme of each trial.

---

[1] http://www.bbci.de/competition/iv/desc_2a.pdf.

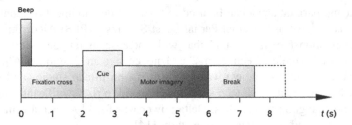

Fig. 2. Timing diagram of one trial

4   Experimental Design

For the experiment, only the first session of the dataset is taken keeping in mind that the next day the participants will get more acquainted with the task and thus will experience less fatigue. The approach to model EEG-fatigue relationship consists of two key steps: a. Extraction of band power of EEG signals and b. Track the growth of fatigue using KPLS.

4.1   Extraction of Band Power of EEG Signals

The first key step is the adoption of a framework that allows us to distinguish two different fatigue states: "*Low fatigue*" and "*High fatigue*". The hypothesis of the experiment is that motor imagery is fatiguing and hence a person experiences fatigue on completion of the motor imagery task. On this basis, we consider the first run as "*Low fatigue state*" and the last run as *High fatigue state*".

As stated in Sect. 2.2, this study adopted two oscillatory EEG rhythms: frontal theta and parietal alpha to analyze mental fatigue in motor imagery. Average power of theta and alpha are measured in Hz at the electrodes Fz and Pz respectively for each of the trial. However, in case of subjects 1, 3, 7, 6 it is found that there is no increase in theta power indicating that these subjects may not experience any fatigue. Hence, we track the growth of fatigue for the subjects 2, 4, 5, 8, and 9.

Figure 3 depicts the marked increase in parietal alpha and frontal theta from first to last runs. The topographic maps in this paper are plotted using topoplot function from Matlab. The figure shows the alpha and theta power for the subjects 2, 4, 5, 8 and 9. The dark red color in the topographic maps indicates the increase in parietal alpha and frontal theta power. The graphs in the third row in the figure show the growth of alpha and theta power during each run. The X axis of the graphs shows the number of runs and the Y axis shows the alpha and theta power respectively. The first graph shows progressive increase of alpha (6.4 to 8 Hz) with peak 8 Hz during last run while the second graph shows marked increase in theta (1 to 6 Hz) with peak 5.8 Hz during last run.

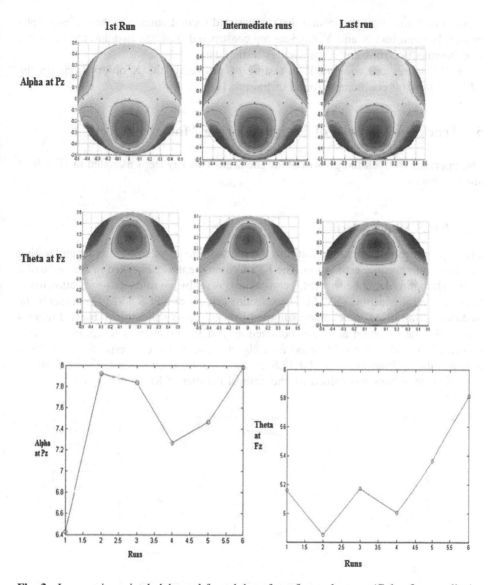

Fig. 3. Increase in parietal alpha and frontal theta from first to last run (Color figure online)

4.2   Kernel Partial Least Squares to Track the Growth of Fatigue

The second step of the approach is to employ Kernel Partial Least Squares (KPLS) [2] to track the growth of fatigue during motor imagery i.e. how it progresses from the low fatigue to high fatigue state. KPLS takes as input two matrices X and Y, X being the set of predictors and Y being the set of response variables. The Y matrix is a vector of −1 or +1 for representing two classes - the low fatigue state and the high fatigue state for training. The X is an $n \times 2$ matrix where n is the number of observations. First column of X matrix is the frontal theta power and second column is the parietal alpha power. For training, X and Y matrices are constructed from the first and the last runs. The intermediate runs are considered as testing data.

KPLS scores are computed[2] in testing by projecting the X observations on the KPLS regression coefficients found in training [1].

5   Track the Growth of Fatigue During Motor Imagery

The approach to track the growth of fatigue needs two key steps as stated in [1] which are as follows:

5.1   KPLS Model Selection

EEG epochs from the first and last runs are splitted into equal sized training and testing partitions for classifier estimation. The KPLS components are checked in the range of 1 to 10. The optimal number of KPLS components is set by ten fold cross-validation using Support Vector Machine (SVM). The selection criteria of KPLS-SVM [3] model is the maximum classification accuracy summed over all cross validation subsets. Figure 4 shows the classification accuracy obtained against the number of KPLS components where S2, S4, S5, S8 and S9 stand for Subjects 2, 4, 5, 8 and 9 respectively. Table 2 portrays the optimal number of KPLS components for each subject. Figure 5 shows classification accuracy obtained for the optimal number of KPLS components.

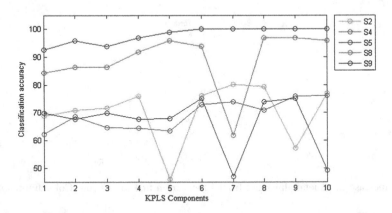

Fig. 4. Classification accuracy against the number of KPLS components

---

[2] The Matlab code for KPLS has been taken from Roman Rosipal's homepage: http://aiolos.um.savba. sk/~roman/soft_data.html.

Table 2. Classification accuracy against optimal number of KPLS components

Subject	Optimal no. of KPLS components
Subject 2	8
Subject 4	9
Subject 5	9
Subject 8	8
Subject 9	6

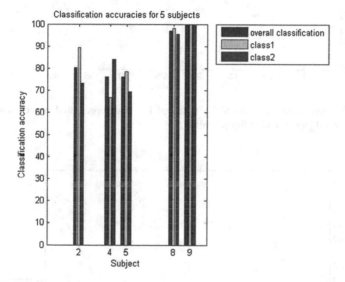

Fig. 5. Classification accuracy against the optimal number of KPLS components

5.2 KPLS Model Prediction

The predictive validity of the KPLS-SVM model is examined by testing them with data from the intermediate runs. The KPLS model selected based on the training data gives KPLS regression coefficients. KPLS scores of the intermediate runs are computed by projecting the test data on the KPLS regression coefficients for each trial. Similarly, KPLS scores for the first and last runs are computed by projecting the training data on the KPLS regression coefficients. Negative KPLS scores reflect the low fatigue state and positive KPLS scores reflect the high fatigue state. Figure 6 depicts that the KPLS scores of first and last run for subject 9. The figure portrays that KPLS scores occupies distinct space for the two classes.

For each of the 5 subjects, for each run, we computed the mean of the KPLS scores and plot the scores in a graph as shown in Fig. 7. The figure portrays the orderly progression from low fatigue to high fatigue state.

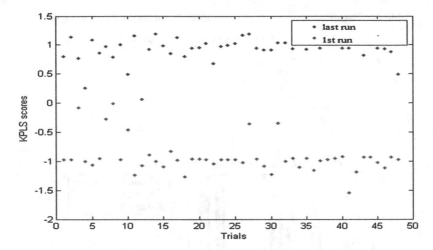

Fig. 6. KPLS scores computed from the first and last run represented by red and blue dots respectively for subject 9 (Color figure online)

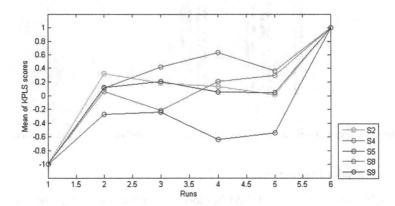

Fig. 7. Means of KPLS scores for each run in case of 5 subjects

6 Final Comments

The KPLS model shows an orderly relationship between the two band power estimates and the two fatigue states. Power in parietal alpha and frontal theta EEG bands across the 5 subjects shows a difference between the first and the last runs. As identified in the literature, mental fatigue is correlated with increase in parietal alpha and frontal theta. The study makes use of only these two parameters to estimate mental fatigue without any validation with other measures like self reports, performance or observed behavior since the dataset has no information about these depicting the level of mental fatigue with time. It is found that Subject 5 shows high fatigue only in case of the last run. Several reasons can be put forward for this kind of disparity. In case of this subject, it is seen that average theta power is greater than the average alpha power which is a sign of BCI illiteracy [15]. Another reason for this kind of disparity may be the incorrect predictive model. Third, in case of the dataset, the motor imagery task was carried out for duration of approx. 45 min. only. This length of a period may not be tedious enough for a person to experience fatigue. Further, breaks are provided after each run and each trial which may decrease the level of fatigue.

Future research direction is to develop a method that stabilizes the link between EEG features and cognitive fatigue over long period of time. Validating the correlation with other measures of fatigue like self reports, performance measures is part of ongoing research. The idea also includes determining which spectral features of EEG leads to more accurate prediction.

Acknowledgement. Financial support from MHRD as Centre of Excellence on Machine Learning Research and Big Data Analysis is acknowledged. Many thanks to Dr. Roman Rosipal and Dr. Leonardo de Trezo for providing us with their much needed help. Assistance received under DST-UKEIRI Project: DST/INT/UK/P-91/2014 is gratefully acknowledged.

References

1. Trejo, L.J., Kubitz, K., Rosipal, R., Kochavi, R.L., Montgomery, L.D.: EEG-based estimation and classification of mental fatigue. Psychology **6**(5), 572–589 (2015)
2. Rosipal, R., Trejo, L.J.: Kernel partial least squares regression in reproducing kernel hilbert space. J. Mach. Learn. Res. **2**, 97–123 (2001)
3. Rosipal, R., Trejo, L.J., Matthews, B.: Kernel PLS-SVC for linear and nonlinear classification. In: ICML, pp. 640–647 (2003)
4. Borghini, G., Astolfi, L., Vecchiato, G., Mattia, D., Babiloni, F.: Measuring neurophysiological signals in aircraft pilots and car drivers for the assessment of mental workload, fatigue and drowsiness. Neurosci. Biobehav. Rev. **44**, 58–75 (2004)
5. Dawson, D., Reid, K.: Fatigue, alcohol and performance impairment. Nature **388**(6639), 235–235 (1997)
6. Pomer-Escher, A., Tello, R., Castillo, J., Bastos-Filho, T.: Analysis of mental fatigue in motor imagery and emotional stimulation based on EEG. In: XXIV Congresso Brasileiro de Engenharia Biomedica-CBEB (2014)
7. Liu, J., Zhang, C., Zheng, C.: EEG-based estimation of mental fatigue by using KPCA–HMM and complexity parameters. Biomed. Sign. Process. Control **5**(2), 124–130 (2010)
8. Holm, A., Lukander, K., Korpela, J., Sallinen, M., Müller, K.M.: Estimating brain load from the EEG. Sci. World J. **9**, 639–651 (2009)
9. Cao, T., Wan, F., Wong, C.M., da Cruz, J.N., Hu, Y.: Objective evaluation of fatigue by EEG spectral analysis in steady-state visual evoked potential-based brain-computer interfaces. Biomed. Eng. online **13**(1), 28–40 (2014)
10. Jap, B.T., Lal, S., Fischer, P., Bekiaris, E.: Using EEG spectral components to assess algorithms for detecting fatigue. Expert Syst. Appl. **36**(2), 2352–2359 (2009)
11. Zarjam, P., Epps, J., Chen, F.: Evaluation of working memory load using EEG signals. In: Proceedings of APSIPA Annual Summit and Conference, pp. 715–719 (2010)
12. Kumar, N., Kumar, J.: Measurement of cognitive load in HCI systems using EEG power spectrum: an experimental study. Procedia Comput. Sci. **84**, 70–78 (2016)

13. Freeman, W.J.: Making sense of brain waves: the most baffling frontier in neuroscience. In: International Conference on Biocomputing, Grainesvile (2001)
14. Craig, A., Tran, Y., Wijesuriya, N., Nguyen, H.: Regional brain wave activity changes associated with fatigue. Psychophysiology 49(4), 574–582 (2012)
15. Ahn, M., Cho, H., Ahn, S., Jun, S.C.: High theta and low alpha powers may be indicative of BCI-illiteracy in motor imagery. PloS one 8(11), 1–11 (2013). e80886
16. Gevins, A., Smith, M.E., Leong, H., McEvoy, L., Whitfield, S., Du, R., Rush, G.: Monitoring working memory load during computer-based tasks with EEG pattern recognition methods. Hum. Factors J. Hum. Factors Ergon. Soc. 40(1), 79–91 (1998)
17. Lin, C.T., Wu, R.C., Liang, S.F., Chao, W.H., Chen, Y.J., Jung, T.P.: EEG-based drowsiness estimation for safety driving using independent component analysis. IEEE Trans. Circ. Syst. I Regul. Pap. 52(12), 2726–2738 (2005)
18. Roy, R.N., Bonnet, S., Charbonnier, S., Campagne, A.: Mental fatigue and working memory load estimation: interaction and implications for EEG-based passive BCI. In: 2013 35th Annual International Conference of the IEEE Engineering in Medicine and Biology Society (EMBC), pp. 6607–6610. IEEE (2013)
19. Kumar, N., Khaund, K., Hazarika, S.M.: Bispectral analysis of EEG for emotion recognition. Procedia Comput. Sci. 84, 31–35 (2016)
20. Koelstra, S., Muhl, C., Soleymani, M., Lee, J.S., Yazdani, A., Ebrahimi, T., Pun, T., Nijholt, A., Patras, I.: DEAP: a database for emotion analysis; using physiological signals. IEEE Trans. Affect. Comput. 3(1), 18–31 (2012)

HCI Applications and Technology

Graph-Based Clustering for Apictorial Jigsaw Puzzles of Hand Shredded Content-less Pages

Lalitha K.S.[1]([⊠]), Sukhendu Das[1], Arun Menon[2], and Koshy Varghese[2]

[1] Department of Computer Science and Engineering, IIT Madras, Chennai, India
kslalitha@cse.iitm.ac.in, sdas@iitm.ac.in
[2] Department of Civil Engineering, IIT Madras, Chennai, India
{arunmenon,koshy}@iitm.ac.in

Abstract. Reassembling hand shredded content-less pages is a challenging task, with applications in forensics and fun games. This paper proposes an efficient iterative framework to solve apictorial jigsaw puzzles of hand shredded content-less pages, using only the shape information. The proposed framework consists of four phases. In the first phase, normalized shape features are extracted from fragment contours. Then, for all possible matches between pairs of fragments transformation parameters for alignment of fragments and three goodness scores are estimated. In the third phase, incorrect matches are eliminated based on the score values. The alignments are refined by pruning the set of pairwise matched fragments. Finally, a modified graph-based framework for agglomerative clustering is used to globally reassemble the page(s). Experimental evaluation of our proposed framework on an annotated dataset of shredded documents shows the efficiency in the reconstruction of multiple content-less pages from arbitrarily torn fragments.

Keywords: Content-less page reassembly · Partial contour matching · Shape features · Agglomerative clustering · Global reassembly

1 Introduction

Paper documents are prone to damages caused by environmental conditions or human activities. When documents are shredded into fragments, they become indecipherable. Converting this indecipherable form to a decipherable form is a task of great importance in fields like forensics and archaeology. The process becomes even more complex, when the shredded fragments have no content in it. Typically, features of fragments based on shape, color, texture or combinations of these, are used in reassembly. However, apictorial puzzles have no texture content in them. Hence, we solve the problem of reassembly using the shape (contour) information alone. In this paper, we propose an approach to automatically reassemble the hand shredded fragments of content-less pages.

In our work, we use the *bdw082010* dataset [11], which has 48 double sided sheets, shredded by hand into 8 fragments each. Since, these sheets have information on both front and back sides, the dataset contains 96 pages. The fragments in this dataset have almost arbitrary shapes. For reassembly, only the

© Springer International Publishing AG 2017
A. Basu et al. (Eds.): IHCI 2016, LNCS 10127, pp. 135–147, 2017.
DOI: 10.1007/978-3-319-52503-7_11

binary segmentation mask of each of the fragments is used. Figure 1(a–b) shows eight fragments (with contents) from a page and the original document itself. Figure 1(c) shows the binary masks of the fragments. Our work reassembles the eight fragments in Fig. 1(c) into Fig. 1(d). Figure 1(e) shows the failure of method proposed in [6], one of the state of art methods to solve apictorial jigsaw puzzle.

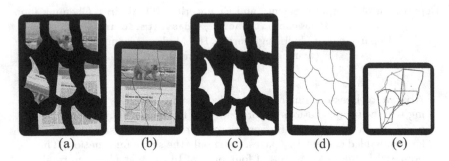

    (a)   (b)   (c)   (d)   (e)

Fig. 1. Example page from dataset [11]. (a) Eight hand shredded paper fragments with content. (b) Reassembled page with content (ground-truth). (c) Binary masks (content-less) of the paper fragments, used as input for our problem. (d) Page Reassembled using the binary masks in (c), by our proposed method. (e) Result generated by the method proposed in [6], when the features are extracted from the smoothed contours of the fragments in (c).

The main contributions of our work are as follows:

1. We propose an efficient framework for the global reassembly of the fragments, which is similar to agglomerative clustering technique.
2. Our approach is unsupervised, that is, it does not require any prior information about the fragments or the final template, and it tends to pack the fragments densely rather than filling the gaps.

The rest of the paper is organized as follows: Sect. 2 discusses the related work. Section 3 gives a detailed overview of the framework proposed for reassembly of hand shredded content-less pages. Section 4 describes the experimental results. Section 5 concludes the work.

2   Related Works

The problem of reassembly of fragmented sheets is solved by considering shape-related features in [4–8,10,16,19,20], color/texture-related features in [14,17] and combination of shape-related & color-related features in [9,11,18]. In some works, as in [16], the contours are divided into contour segments, delimited by the corners, which are matched. Generally, distance metrics are used to measure the similarity of the features of the boundary points. In [11], Richter *et al.* proposed a supervised approach using an SVM classifier, to find points in the boundary with similar features. Most of the recent methods rely on texture features in the fragments to align globally and locally.

Global reassembly algorithms use a greedy approach to iteratively merge fragments. In [2], methods for global reassembly includes global relaxation to reduce the error in the final reconstruction. In [18], a graph-based algorithm that performs better groupwise matching and a variational graph optimization to minimize the error in final reconstruction are proposed. In [9], Liu *et al.* proposed a method to reassemble multiple photos simultaneously.

The algorithms developed in [4–6,8,10,20] solve apictorial jigsaw puzzles with specific non-random patterns ("indents" and "outdents") in a fragment's contour. In [5], the "indents" and "outdents" are matched using ellipses fitted into them. In [10], the curve fitting is done using polar coordinate systems centered around the local extrema of the curvature. In [8], dynamic programming methods are used to match the curvature and arc length invariants. Once the matches are known, the robust relative transformations between the fragments are estimated using variants of ICP [13] or variants of MLESAC [12]. Incorrect matches are eliminated by locally verifying the transformation between fragments as in [12] or by checking the global compatibility between matches as in [18].

In [19], the candidate disambiguation problem has been formulated to define the compatibility between neighboring matches and global consistency is defined as the global criterion. Results show the capability of the approach to reassemble two documents with up to 50 pieces. Recent work in [6], efficiently solves apictorial jigsaw puzzles. Here, the contours are decomposed into smaller arcs by *bivertex decomposition* and the arcs are then matched to fit the fragments. Also, an efficient algorithm to minimize the local error in alignment of fragments is proposed. However, as shown in Fig. 1(e), the method does not reassemble the hand shredded pages, as this method relies only on curvature of "indents" and "outdents" in fragments. Additionally, they do not consider the overlap between fragments and the global placement of fragments during the reassembly. Our paper bridges the gap by reassembling shredded content-less pages.

3   Proposed Framework

The main aim of the paper is to reassemble a hand shredded content-less page using shape information alone. The proposed iterative framework, as shown in Fig. 2, suggests a way for the reassembly. The stages are discussed below:

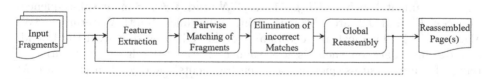

Fig. 2. The proposed iterative framework with various processing stages, for hand shredded content-less page reassembly.

3.1 Feature Extraction

To reduce the complexity of processing, we find the polygonal approximation of the contours of the fragments using the Douglas-Peucker algorithm [3]. However, the number of points in the approximated polygon depends solely on the parameter σ used in this algorithm [3]. Given N input fragments from the shredded page to be reassembled, the sets of input fragment contours and approximated fragment contours be $\widehat{Q} = \{Q^1, Q^2, \ldots, Q^N\}$ and $\widehat{P} = \{P^1, P^2, \ldots, P^N\}$ respectively. Here, $P^i = \{p_1^i, p_2^i, \ldots, p_{n_i}^i\} \subseteq Q^i$ describes the ith fragment, with n_i being the number of points in the approximated i^{th} contour polygon.

Three features are extracted from every vertex of the approximated contour polygon. They are:

- *Signed Curvature* (κ) - assume, a fragment's boundary is of class C^2
- *Mean of Edge Lengths* (μ), from current vertex to the two adjacent vertices
- *Angle between two vectors* (θ), one joining previous vertex & current vertex and other joining current vertex & next vertex

Thus, the feature representation, $f = [\kappa, \mu, \theta]$, at a vertex is a 3-dimensional vector. In some cases, features have a skewed distribution. To overcome this, skewness of every feature distribution is obtained and appropriate mathematical operations are performed on the feature values to reduce the skewness. Finally, these features are combined to form a representative feature vector. The new representative feature is defined as, $\hat{f} = [\log(|\kappa|), \log(\mu), \theta]$. Features are then normalized using *Min-Max Normalization*. Let, $F_c^i, F_a^i \in \mathbb{R}^{n_i \times 3}$ be the matrices containing the normalized feature vectors of the ith fragment, when the points in the polygon are traversed in clockwise and anti-clockwise direction, respectively.

3.2 Pairwise Matching of Fragments

This phase discovers the set of all possible matches between two fragments. Based on the matches, a relative transformation necessary to align one fragment with another is estimated. For each transformation, scores are computed based on the goodness of the alignment. For N input fragments, there are $^N C_2$ possible pairings, for which the process is repeated.

Discovery of Contour Segment Matches. Let i and j be two randomly chosen fragments, such that $i, j \in 1, 2, \ldots, N$ and $i \neq j$. Similar sub-sequences between F_c^i and F_a^j are obtained using a modified Smith-Waterman (SW) algorithm [15]. Two feature vectors are considered to be similar, if the Euclidean distance between them is less than ζ. If the SW algorithm returns M pairs of common sub-sequences between fragments i and j, then there are M contour segments in i and j that are matching. If sp_m^i and ep_m^i denote the indices of starting and ending points of the contour segment in fragment i corresponding to m^{th} match, then $cs_m^i = \{p_{sp_m^i}^i, \ldots, p_{ep_m^i}^i\} \subseteq P^i$ denotes the sequence of points in fragment i for the mth match, $m \in 1, \ldots, M$. Thus, the set

$\mathbf{CS}^{i,j} = \{\{cs_1^i, cs_1^j\}, \{cs_2^i, cs_2^j\}, \ldots, \{cs_M^i, cs_M^j\}\}$ contains all the matched contour segments pairs. Matching using F_a^i and F_c^j also gives rise to the same set of matches, but the ordering (traversal direction) of points in the contour segments are reversed.

Estimation of Transformation and Scores. For every matched contour segments pair in $\mathbf{CS}^{i,j}$, a transformation is estimated. Unlike the prior works [12,13], we find a pair of points $\alpha, \alpha' \in cs_m^i$ and the corresponding points $\beta, \beta' \in cs_m^j$, such that $|\text{Len}(\alpha, \alpha') - \text{Len}(\beta, \beta')| \leq \delta$, where $\text{Len}(\alpha, \alpha')$ is the arc length between α and α', and δ is a parameter. A translation vector, $\boldsymbol{\lambda}$, is computed from the offset between α and β, as $\boldsymbol{\lambda} = \alpha - \beta$. The rotation angle, ϕ, is computed by computing the angle between vectors $\overrightarrow{\alpha\alpha'}$ and $\overrightarrow{\beta\beta'}$. The Euclidean transformation, $T_m^{i,j} \in \text{SE}(2)$, applied to fragment j to align it with fragment i, is thus represented using its parameters as $T_m^{i,j} = (\phi, \boldsymbol{\lambda}) \in S^1 \times \mathbb{R}^2$ and the set containing the transformations between i and j is $\mathbf{T}^{i,j} = \{T_1^{i,j}, T_2^{i,j}, \ldots, T_M^{i,j}\}$. For each $T_m^{i,j} \in \mathbf{T}^{i,j}$, we calculate three scores, which are commonly used as topological features for 2-D shapes:

1. *Connectivity between fragments* is the ratio of the length of the common sub-sequence between the fragments to the minimum of the two perimeters.

$$CON_m^{i,j} = \frac{\text{Len}(cs_m^i)}{\min(\text{Perimeter}(P^i), \text{Perimeter}(P^j))}, \begin{cases} i, j \in 1, \ldots, N, \ i \neq j, \\ m = 1, \ldots, M. \end{cases} \quad (1)$$

2. *Relative Fitness value* is the ratio of the sum of the areas of overlap & gap between fragments to the mean of the areas of the fragments.

$$FIT_m^{i,j} = 2\frac{|OA^{i,j}| + |GA^{i,j}|}{\text{Area}(P^i) + \text{Area}(P^j)}, \begin{cases} i, j \in 1, \ldots, N, \ i \neq j, \\ m = 1, \ldots, M. \end{cases},$$

where, $OA^{i,j} = \text{Area}(P^i \cap TP^j)$,

$$GA^{i,j} = \text{Area}(P_m^{i,j}) - \text{Area}(P^i \cup TP^j),$$

$$P_m^{i,j} = \text{Boundary}(P^i \cup TP^j), \text{ where } TP^j = T_m^{i,j} \star P^j. \quad (2)$$

Operator '$\star$' denotes the transformation for all points in a fragment.

3. *Compactness of the merged Fragment* is the ratio of the square of the perimeter of the merged fragment to the area of the merged fragment.

$$COM_m^{i,j} = \frac{[\text{Perimeter}(P_m^{i,j})]^2}{\text{Area}(P_m^{i,j})}, \begin{cases} i, j \in 1, \ldots, N, \ i \neq j, \\ m = 1, \ldots, M. \end{cases} \quad (3)$$

These scores penalize overlap between a pair of fragments, and also ensure that the fragments are densely packed.

3.3   Elimination of Incorrect Matches and Graph Generation

This phase eliminates the incorrect matches between two fragments based on the associated scores and finds the best possible alignment between two fragments. This is a four step process. First, we eliminate the matches that have *Connectivity* scores less than th_{CON} and *Fitness* scores greater than th_{FIT}, and the remaining transformations form a set of $\widetilde{\mathbf{T}}^{i,j} \subseteq \mathbf{T}^{i,j}$, $\forall i, j \in 1, \ldots, N$ and $i \neq j$, such that the cardinality of the set is $\widetilde{\widetilde{\mathbf{T}}}^{i,j} = M' \leq M$.

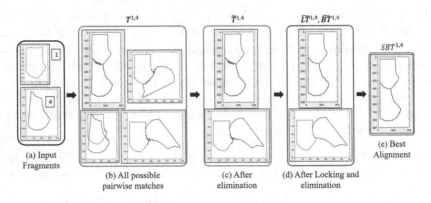

(a) Input Fragments

(b) All possible pairwise matches

(c) After elimination

(d) After Locking and elimination

(e) Best Alignment

Fig. 3. Elimination of incorrect matches between fragments 1 and 4, for the example shown in Fig. 1(c). The fragment indices are shown in rectangular boxes. (a) Pair of input fragment contours 1 and 4. (b) All possible pairwise matches between the fragments. (c) Pruned matches after elimination, based on transformation scores (Eq. (1), (2) and (3)). (d) Matches retained after locking the fragments and again eliminating matches based on transformation scores. (e) Best alignment between the pair.

Then, the Locking algorithm proposed in [6] is used to reduce the errors in the transformations. The parameters values of $K_1, \ldots, K_4$, ϵ, ν, ρ and j_{max} used are same as that given in [6]. Let, $\widetilde{\mathbf{LT}}^{i,j} = \text{Locking}(\widetilde{\mathbf{T}}^{i,j})$, $\widetilde{\widetilde{\mathbf{LT}}}^{i,j} = M'$, be the set of error-corrected transformations. Application of this algorithm leads to increase in the *Connectivity* score and decrease in the *Fitness* score (recomputed).

Then, based on the *Connectivity* score, we again eliminate transformations and form a set $\widetilde{\mathbf{BT}}^{i,j} = \{\widetilde{BT}_1^{i,j}, \ldots, \widetilde{BT}_{M''}^{i,j}\} \subseteq \widetilde{\mathbf{LT}}^{i,j}$ and $\widetilde{\widetilde{\mathbf{BT}}}^{i,j} = M'' \leq M'$.

Finally, Single Best Transformation that align j with i, $SBT^{i,j}$, is estimated based on the *Compactness* (Cmpts) score of the transformations in $\widetilde{\mathbf{BT}}^{i,j}$ as:

$$SBT^{i,j} = \begin{cases} \underset{\substack{\widetilde{BT}_k^{i,j} \\ k=1,2,\ldots,M''}}{\arg\min} \ \text{Cmpts}(\widetilde{BT}_k^{i,j}), & \text{if } \widetilde{\widetilde{\mathbf{BT}}}^{i,j} \neq 0, \\ \\ null, & \text{Otherwise} \end{cases} \quad \begin{array}{l} \forall i, j \in 1, \ldots, N, \\ i \neq j. \end{array} \quad (4)$$

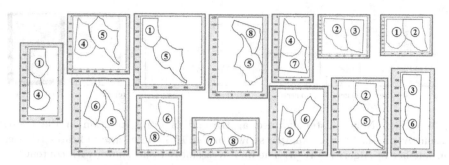

Fig. 4. All pairwise matches corresponding to the transformations in the set **SBT**, for the example in Fig. 1(c). Pairwise matches with green border are true matches, and one with red border is a false match. Fragment indices are given in circles. (Color figure online)

From Eq. (4), the single best transformation to align i with j is:

$$SBT^{j,i} = \begin{cases} (SBT^{i,j})^{-1}, & \text{if } SBT^{i,j} \neq null \\ null, & \text{Otherwise} \end{cases}, \qquad \begin{matrix} \forall i, j \in 1, \ldots, N, \\ i \neq j. \end{matrix} \qquad (5)$$

The above two transformations are appended to the set **SBT**, containing all the Single Best Transformations, $\mathbf{SBT} = \mathbf{SBT} \cup \{SBT^{i,j}, SBT^{j,i}\}$. If $SBT^{i,j} \neq null$ and $SBT^{j,i} \neq null$, we then add an edge, e_{ij}, between nodes i and j in the undirected graph, $G(\mathbf{V}, \mathbf{E})$, formed with input fragments as its nodes. Weight of the edge e_{ij} is:

$$w(e_{ij}) = \text{Cmpts}(SBT^{i,j}), \qquad i, j \in 1, \ldots, N, \ i \neq j. \qquad (6)$$

Figure 3(a)–(e) shows an example of the process of elimination of incorrect pairwise matches of fragments. The final match in Fig. 3(e) is the best of all the matches identified by the Pairwise Fragment Matching phase in Fig. 3(b). Locking reduces the error in the alignment. The above steps are applied for all the possible pairings. Figure 4 shows matches corresponding to the transformations in the set **SBT**, for the example in Fig. 1(c).

3.4   Global Reassembly

In this phase, transformations are applied to approximately align the fragments. Unlike other methods that start from a seed fragment and then greedily aligns other fragments, the method proposed in this paper follows a Modified Agglomerative Clustering Algorithm to align the fragments together. Initially, consider each fragment as a cluster. Then, detect clusters corresponding to the best alignment, based on the edge weights of the graph G. Here, lesser the edge weight better is the alignment. Merge the pair of clusters, if the alignment of fragments in the participating clusters do not lead to overlap of fragments in the reassembled page. Choose the next best alignment and repeat the steps until there are no more clusters that can be merged without overlapping.

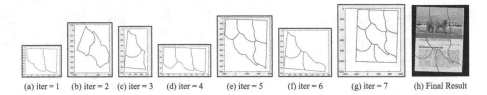

(a) iter = 1   (b) iter = 2   (c) iter = 3   (d) iter = 4   (e) iter = 5   (f) iter = 6   (g) iter = 7   (h) Final Result

Fig. 5. (a)–(g) New Clusters formed at the end of every iteration in the Global Reassembly phase, for the example shown in Fig. 1(c). Here, *iter* indicates the iteration number in Algorithm 1. (h) The final reassembly result shown with content.

When fragments are clustered together, it implies that fragments can be assembled without any overlap, by applying the appropriate transformations. The proposed method for global reassembly is similar to an algorithm proposed in [11]. However, in the proposed method, the weight matrix need not be updated after every iteration. This reduces the computational overhead significantly.

The steps of the proposed method for global reassembly are given in Algorithm 1. We start by considering N singleton clusters as an initial set of clusters $\mathcal{C}^{(0)} = \{c_1, \ldots, c_N\}$. Each cluster contains a fragment index, $c_i = \{i\}, i \in 1, \ldots, N$. The transformations to be applied to the fragments corresponding to the elements of cluster c_k is stored in the set $\tau_k \in \text{SE}(2)$. Initialize $\tau_k = \{I\}, \forall k = 1, \ldots, N$, where $I = (0,0,0) \in S^1 \times \mathbb{R}^2$ represents the identity transformation. Define a set containing all the initial transformations of the clusters as $\Gamma^{(0)} = \{\tau_1, \ldots, \tau_N\}$. The sets $\mathcal{C}^{(0)}$ and $\Gamma^{(0)}$, along with set of fragments $(\widehat{\mathcal{Q}})$, graph (G) and set of single best transformations (**SBT**), are input to Algorithm 1.

Algorithm 2 shows the steps of the function used to combine clusters and to compute the transformations for elements in the new cluster. Figure 5 shows examples of new valid clusters formed at the end of each iteration of Algorithm 1. Figure 5(h) shows the page reassembled by the proposed method.

4   Experiments and Results

Experiments are done to evaluate and verify performance of methods in real world scenarios. The proposed method is capable of reassembling all the 96 pages (front and back sides of 48 sheets), within 2 iterations.

The parameters used in all the experiments are empirically set, as follows:

$$\sigma : 1.4\text{--}3.1, \qquad \zeta : 0.06\text{--}0.10, \qquad \epsilon = 0.0001, \qquad \nu = 3,$$
$$K_1 = 15, \qquad K_2 = 4, \qquad K_3 = 1, \qquad K_4 = 0.5, \qquad \delta = 5,$$
$$th_{CON} : 0.10\text{--}0.15, \qquad th_{FIT} : 0.009\text{--}0.03, \qquad \rho = 1/3, \qquad j_{max} = 50.$$

The performance of the proposed algorithm solely depends on the choice of the above parameters. The efficiency of using $\hat{f} = [\log(|\kappa|), \log(\mu), \theta]$ as the feature vector at a vertex, instead of using $f = [\kappa, \mu, \theta]$ in the feature extraction phase,

Algorithm 1. *Algorithm for Global Reassembly:* Initial set of Fragments are given as input. Output contains clusters of fragments, that are aligned together without overlap, and their transformations.

Input: $\widehat{\mathcal{Q}} = \{Q^1, \ldots, Q^N\}$: Set of fragments, $G(\mathbf{V}, \mathbf{E})$: Graph,
$\quad\quad\, \mathcal{C}^{(0)}$: Initial set of clusters, $\Gamma^{(0)}$: Initial set of cluster transformations,
$\quad\quad\, \mathbf{SBT}$: Set of Single Best Transformations

Output: $\mathcal{C}^{(size(\mathbf{E}))}$: Final set of clusters,
$\quad\quad\quad\, \Gamma^{(size(\mathbf{E}))}$: Final set of cluster transformations

---

1 Sort $\mathbf{E}$ in increasing order of weights
2 Set $iter \leftarrow 1$
3 **while** $iter \leq size(\mathbf{E})$ **do**
4　　Assign u and v to the nodes connected by edge $\mathbf{E}(iter)$
5　　Find the clusters c_U and c_V to which u and v belong to, respectively
6　　Set $flag \leftarrow 0$
7　　**if** c_U and c_V are not same **then**
8　　　$(\widehat{c_U}, \widehat{\tau_U}) \leftarrow \mathtt{NewCluster}(u, v, c_U, c_V, \tau_U, \tau_V, SBT^{u,v})$ /* See Algorithm 2 */
9　　　**if** fragments in $\widehat{c_U}$ do not overlap **then**
10　　　　$\mathcal{C}^{(t)} = \{\mathcal{C}^{(t-1)} \setminus \{c_U, c_V\}\} \cup \widehat{c_U};\ \Gamma^{(t)} = \{\Gamma^{(t-1)} \setminus \{\tau_U, \tau_V\}\} \cup \widehat{\tau_U}$
11　　　　Set $flag \leftarrow 1$
12　　　**end**
13　　**end**
14　　**if** $flag - 0$ **then**
15　　　$\mathcal{C}^{(t)} = \mathcal{C}^{(t-1)};\ \Gamma^{(t)} = \Gamma^{(t-1)}$
16　　**end**
17　　Increment $iter$ by 1
18 **end**
19 **return** $\left(\mathcal{C}^{(size(\mathbf{E}))},\ \Gamma^{(size(\mathbf{E}))}\right)$

---

is shown by computing the *False Discovery Rate* (FDR). Here, FDR is the ratio of the number of false matches to the total number of matches in the set **SBT**. Table 1 gives the values of FDR for three sheets in the dataset, using f and $\hat{f}$ as feature vectors at a vertex. Using $\hat{f}$ as feature vector helps in finding more true matches and simultaneously reducing the number of false matches.

Results in Fig. 6 show that the *bivertex decomposition* method, proposed in [6] to find the initial set of possible matches, returns a larger set of hypotheses than our proposed method. Thus, the running time of the proposed method is less. According to [1], the worst-case lower bound on the number of iterations performed by ICP algorithm in order to converge is $\Omega(n/d)^{d+1}$, where n is the size of the input data point set and d is the dimensionality of the input data. In this work $d = 2$ and thus, if ICP algorithm is used to estimate the transformation, given n pairs of matched points, then the worst-case lower bound on the number of iterations performed in order to converge is $\Omega(n^3)$. However, the running time complexity of the method proposed in this paper, in Pairwise Matching of

Algorithm 2. *Function to Combine Clusters:* Two clusters (c_U, c_V) along with their transformations (τ_U, τ_V) are the input. Output is a new cluster $(\widehat{c_U})$ formed by combining the input clusters and transformations $(\widehat{\tau_U})$ to be applied to the elements in the new cluster.

```
1  Function NewCluster(u, v, c_U, c_V, τ_U, τ_V, SBT^{u,v})
2  |   New cluster, ĉ_U = {c_U ∪ {v} ∪ {c_V \ {v}}}
3  |   Initialize τ̂_U = ∅
4  |   for each l in c_U do
5  |   |   τ̂_U = τ̂_U ∪ {tran(U, l)}
   |   |   /* tran(U, l) ∈ τ_U represents the transformation applied to
   |   |      fragment corresponding to element l in cluster c_U        */
6  |   end
7  |   Append transformation for element v as τ̂_U = τ̂_U ∪ {tran(U, u) * SBT^{u,v}}
   |   /* Operator '*' denotes multiplication of transformations        */
8  |   temp = (tran(U, u) * SBT^{u,v}) * Inverse(tran(V, v))
9  |   for each l in {c_V \ {v}} do
10 |   |   τ̂_U = τ̂_U ∪ {temp * tran(V, l)}
11 |   end
12 return (ĉ_U, τ̂_U)
```

Table 1. FDR values using f and $\hat{f}$ as feature vectors at a vertex of the approximated contour polygons of the input fragments.

Sheet index	Using f	Using $\hat{f}$
Sheet 1 - back	0.125 (=1/8)	0.077 (=1/13)
Sheet 17 - back	0.000 (=0/3)	0.010 (=1/10)
Sheet 44 - front	0.555 (=5/9)	0.222 (=2/9)

Fragments phase, to estimate the transformation is $O(n^2)$. Hence, the proposed work is much faster than the prior work, in estimating the transformation.

Since, the fragments are content-less, it is hard to find which side (front or back) of the fragment should be used in reassembly. To evaluate the performance in such scenarios, both sides of the scanned fragments are given as input. Figure 7 shows the reconstruction of one such sheet. It is observed that the method is capable of reconstructing both front and back sides of the sheet simultaneously.

When fragments are from multiple pages and the number of pages that has been shredded is unknown, the problem gets complicated. Figure 8(b) shows the result of the proposed method, when the input fragments, shown in Fig. 8(a), are from two different pages. Figure 9(b) shows the result of one failure case of the proposed method, for the input in Fig. 9(a). The failure is due to an error in matching of fragments.

Number of initial posssible matches discovered

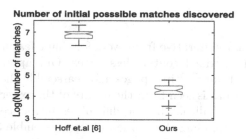

Fig. 6. Number of initial possible matches discovered. Box plots show the median values (red line) as well as the 25th and 75th percentiles. The '+' symbols indicate the outliers. (Color figure online)

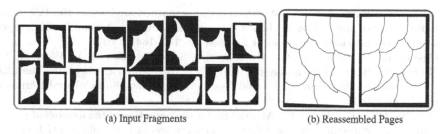

(a) Input Fragments   (b) Reassembled Pages

Fig. 7. Simultaneous reassembly of front and back of Sheet 1, from dataset [11]. (a) Input: 8 fragments, both front and back sides scans. (b) Reassembled pages.

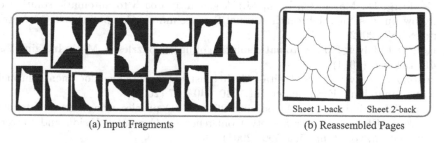

(a) Input Fragments   (b) Reassembled Pages

Fig. 8. Simultaneous reassembly of two pages (back side of Sheet 1 and Sheet 2), from dataset [11]. (a) 16 input fragments. (b) Two correctly reassembled Pages.

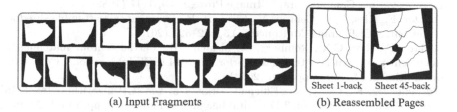

(a) Input Fragments   (b) Reassembled Pages

Fig. 9. Simultaneous reassembly of two pages (back side of Sheet 1 and Sheet 45), from dataset [11]. (a) 16 input fragments. (b) Two reassembled Pages, with one failure case.

146 L.K.S. et al.

5 Conclusion

We have proposed a novel iterative framework for shape feature based automatic reassembly of hand shredded content-less pages. Our approach is capable of processing fragments from multiple pages and reassemble them simultaneously. The efficiency of our work is seen from the results of the experiments performed.

In future work, we shall handle the difficulties introduced due to material loss. Exploration for extending the framework to reassemble 3D broken objects would be an appropriate scope of future work.

References

1. Arthur, D., Vassilvitskii, S.: Worst-case and smoothed analysis of the ICP algorithm, with an application to the k-means method. In: 47th Annual IEEE Symposium on Foundations of Computer Science, pp. 153–164 (2006)
2. Castañeda, A.G., Brown, B.J., Rusinkiewicz, S., Funkhouser, T.A., Weyrich, T.: Global consistency in the automatic assembly of fragmented artefacts. In: The 12th International Symposium on Virtual Reality, Archaeology and Cultural Heritage, pp. 73–80 (2011)
3. Douglas, D.H., Peucker, T.K.: Algorithms for the reduction of the number of points required to represent a digitized line or its caricature. Cartographica Int. J. Geogr. Inf. Geovisualization **10**, 112–122 (1973)
4. Freeman, H., Garder, L.: Apictorial jigsaw puzzles: the computer solution of a problem in pattern recognition. IEEE Trans. Electron. Comput. **13**, 118–127 (1964)
5. Goldberg, D., Malon, C., Bern, M.: A global approach to automatic solution of jigsaw puzzles. In: Eighteenth Annual Symposium on Computational Geometry, pp. 82–87 (2002)
6. Hoff, D.J., Olver, P.J.: Automatic solution of jigsaw puzzles. J. Math. Imaging Vis. **49**, 234–250 (2014)
7. Justino, E., Oliveira, L.S., Freitas, C.: Reconstructing shredded documents through feature matching. Forensic Sci. Int. **160**, 140–147 (2006)
8. Kong, W., Kimia, B.B.: On Solving 2D and 3D puzzles using curve matching. In: 2001 IEEE Computer Society Conference on Computer Vision and Pattern Recognition, vol. 2, pp. 583–590 (2001)
9. Liu, H., Cao, S., Yan, S.: Automated assembly of shredded pieces from multiple photos. IEEE Trans. Multimedia **13**, 1154–1162 (2011)
10. Radack, G.M., Badler, N.I.: Jigsaw puzzle matching using a boundary-centered polar encoding. Comput. Graph. Image Process. **19**, 1–17 (1982)
11. Richter, F., Ries, C.X., Cebron, N., Lienhart, R.: Learning to reassemble shredded documents. IEEE Trans. Multimedia **15**, 582–593 (2013)
12. Richter, F., Ries, C.X., Romberg, S., Lienhart, R.: Partial contour matching for document pieces with content-based prior. In: 2014 IEEE International Conference on Multimedia & Expo, pp. 1–6 (2014)
13. Rusinkiewicz, S., Levoy, M.: Efficient variants of the ICP algorithm. In: Third International Conference on 3-D Digital Imaging and Modeling, pp. 145–152 (2001)
14. Sağiroğlu, M., Erçil, A.: A texture based matching approach for automated assembly of puzzles. In: The 18th International Conference on Pattern Recognition, vol. 3, pp. 1036–1041 (2006)

15. Smith, T.F., Waterman, M.S.: Identification of common molecular subsequences. J. Mol. Biol. **147**, 195–197 (1981)
16. Stieber, A., Schneider, J., Nickolay, B., Krüger, J.: A contour matching algorithm to reconstruct ruptured documents. In: Goesele, M., Roth, S., Kuijper, A., Schiele, B., Schindler, K. (eds.) DAGM 2010. LNCS, vol. 6376, pp. 121–130. Springer, Heidelberg (2010). doi:10.1007/978-3-642-15986-2_13
17. Tsamoura, E., Pitas, I.: Automatic color based reassembly of fragmented images and paintings. IEEE Trans. Image Process. **19**, 680–690 (2010)
18. Zhang, K., Li, X.: A graph-based optimization algorithm for fragmented image reassembly. Graph. Models **76**, 484–495 (2014)
19. Zhu, L., Zhou, Z., Hu, D.: Globally consistent reconstruction of ripped-up documents. IEEE Trans. Pattern Anal. Mach. Intell. **30**, 1–13 (2008)
20. Zisserman, A., Forsyth, D.A., Mundy, J.L., Rothwell, C.A.: Recognizing general curved objects efficiently. In: Geometric Invariance in Computer Vision, pp. 228–251 (1992)

2D Hand Gesture for Numeric Devnagari Sign Language Analyzer Based on Two Cameras

Jayshree Pansare[(✉)] and Maya Ingle

School of Computer Science and Information Technology,
Devi Ahilya Vishwavidyalaya, Indore, India
jayshree.pansare23@gmail.com, mayaingle22@gmail.com

Abstract. Devnagari Sign Language (DSL) has proved to be a powerful and conventional augmentative communication tool especially for speech and hearing impaired. A set of Devnagari script along with DSL numbers is indispensable for understanding Devnagari script via computer system. Proposed Numeric DSL Analyzer (N-DSLA) is designed for the recognition of DSL numbers from "०" to "९" and conversion into Devnagari script along with speech. The architecture of N-DSLA system is fragmented into six consequent phases namely; image capturing, image pre-processing, region extraction, feature extraction, feature matching and pattern recognition. N-DSLA applies Ex-Temp-Match algorithm and Single Hand Two Cameras approach. Our N-DSLA system achieves the recognition rate of 97.2% with recognition time of 0.3 s in complex background with mixed lightning condition. N-DSLA system is employed for handling Music Player application using 10 gestures from "०" to "९" for corresponding operations.

Keywords: Devnagari Sign Language (DSL) · Devnagari numbers · Single hand two camera approach · Ex-Temp-Match algorithm

1 Introduction

Hand Gesture Recognition System (HGRS) for detection of DSL numbers has become essential tool for specific end users (i.e. hearing and speech impaired) to interact with general users via computer system. Numerous HGRS have been developed for recognition of alphanumeric sign languages. Numerous Hand Gesture Recognition System (HGRS) have been developed for recognition of hand gestures using various techniques for varied applications. Wang et al. [1] proposed a real-time gesture detection algorithm for recognizing "incomplete gestures". Modified hidden markov model (HMM) is used for recognizing intentional gestures. The real-time application such as simulated smart board, recognizing lecture activities etc. are developed using detection algorithm by prediction. It accomplishes high recall rate of 93.5% and precision performance of 89.4% for button gesture as well as 84.6% recall rate and 90.7% precision performance for hand gesture. Zhang et al. [2] presented a framework for hand gesture recognition based on accelerometer (ACC) and electromyography (EMG). It recognizes 72 Chinese Sign (CSL) Language words and 40 CSL sentences with accuracy of 95.3% and 96.3% for ACC and EMG respectively. Kim et al. [3] proposed

© Springer International Publishing AG 2017
A. Basu et al. (Eds.): IHCI 2016, LNCS 10127, pp. 148–160, 2017.
DOI: 10.1007/978-3-319-52503-7_12

a novel method tensor canonical correlation analysis to measure video-to-video similarity. The method requires low time complexity and does not require major tuning parameters. Alon, et al. [4] presented a system that recognizes American Sign Language (ASL) digits in cluttered background using vision-based, gesture spotting algorithm. The algorithm increases the detection rate 10 fold, from 8.5% to 85%. Dardas et al. [5] presented a real-time system for interaction with an application via hang gestures. The system includes detection and tracking of bare hand using skin color detection, posture contour comparison algorithm, posture recognition using bag-of-features and multiclass support vector machine (SVM). The system achieves recognition rate of 96.23% under variable scale, orientation, and illumination conditions in cluttered background. In [6] Mohandes et al. reviewed image-based and sensor-based approaches focusing mainly on Arabic Sign Language (ArSL). Kosmidou et al. [7] proposed enhanced sign language recognition using hybrid adaptive weighing process applied to surface electromyogram and 3-D accelerometer data. The system recognizes 61 Greek Sign Language (GSL) signs with highest recognition rate of 97.08%. Yang et al. [8] proposed a novel method for designing threshold models in a conditional random field for distinguishing between sign and nonsign patterns in a vocabulary. The system recognizes signs from isolated data with a recognition rate of 93.5%. Zhu et al. [9] addressed natural human-robot interaction in a smart assisted living system for the elderly and the disabled using neural network and hierarchical HMM. Tindale et al. [10] presented system that captures the gestures of music performers and maps to sounds or used for music transcription. The system applies surrogate sensors for machine learning model for indirect acquisition. Reale et al. [11] presented a vision-based human-computer interaction system to integrate control components using multiple gestures. The system uses two cameras for detection of face and two-camera hand pointing algorithms. Similarly, Tran et al. [12] presented a novel gesture recognition approach for human computer interactivity based on marker-less upper body pose tracking in 3-D with multiple cameras. Yanik et al. [13] developed a system for human-machine interfaces for ubiquitous computing and assistive robotics using growing neural gas algorithm. In [14], Wu et al. have been proposed a novel approach to generate Chinese Taiwanese Sign Language (TSL) based on a predictive sentence template (PST) tree and trigger pair category. In [15], Wang et al. constructed a model for finding similarity between CSL videos using volume local binary patterns. Li et al. [16] identified constituent components of sign gesture and has been improved performance of recognition of sign language using accelerometer and surface electromyography sensors. The system accomplished classification accuracy of 96.5% for 120 Chinese signs. In [17], Zaki et al. has been presented a combination of vision based features for enhancing the recognition of ASL signs using principal component analysis and hidden markov model (HMM). Chen et al. [18] have been designed HGRS for recognition of continuous gestures in stationery background using four modules such as a hand tracking, feature extraction, HMM training, and gesture recognition. The system recognizes 20 gestures with recognition rate of 90%. In [19], Al-Jarrah et al. developed a system to translate manual 30 alphabet gestures in ArSL without using gloves or visual markings. The system identifies the fuzzy inference using subtractive clustering algorithm and least square estimator and applies hybrid learning algorithm for training and achieves 93.55% accuracy. Marrnins et al. [20] presented technologies for the communication between deaf people on e-learning platforms using sign languages. Pansare et al. [21] developed real-time HGRS that recognizes 26 American Sign Language (ASL) alphabets using centroid of Binary Linked Objects (BLOB). This system achieves 90.19% recognition rate in complex background. In [22], Mitra et al. studied and highlighted gesture recognition techniques such as HMM, particle filtering, condensation, skin color etc. In addition to the theoretical aspects, there have been numerous approaches and techniques for recognition of multiple sign languages such as ASL, ArSL, CSL, PSL etc. However, there is an extensive scope to design and develop HGRS for recognition of DSL for various uses. This system will be highly useful to understand the DSL sign postures and for playing music player application. It is evident that many HGRS have been developed for recognition of numbers based on Single Hand Single Camera approach. However, there exists a wide scope to enhance the system design using Single Hand Two Camera approach for auditory handicapped users to interact with general users.

With the potential use of technology, it has become easy to develop the system as per the requirements of varied users including auditory handicapped users. A set of Devnagari numbers ranging from "०" to "९" is indispensable for understanding Devnagari script via computer system. Nexus of our work is on development of the proposed system N-DSLA i.e. Numeric Devnagari Sign Language Analyzer. N-DSLA is based on *Single Hand Two Camera* approach as well as it is used for handling operations of music player application. We discuss the structure of Devnagari script and DSL numbers both in Sect. 2. The detailed system architecture and working of N-DSLA are presented in Sect. 3. Experimental setup of N-DSLA is discussed in Sect. 4. Comparative performance along with experimental results of N-DSLA is entailed in Sect. 5. Lastly, we conclude with conclusion in Sect. 6. Your contribution should be prepared in Microsoft Word.

2   Devnagari Script and DSL Numbers

Devnagari script is also known as Nagari script that covers Devnagari letters and numerals in its set. It is written from left to right and may be given a strong preference for symmetrical rounded shapes within squared outlines from the perspective of its recognition. It is recognizable by a horizontal line that runs along the top of Devnagari letters. Devnagari script has been used for over 120 languages including Hindi, Marathi, Nepali, Pali, Konkani, Bodo, Sindhi and Maithili among other languages and dialects. Therefore, it is considered as most widely used and adopted writing system in the world. Devnagari script consists of 10 primary numbers ranging from "०" to "९" as depicted in Table 1.

Devnagari Sign Language has been proved to be a powerful and conventional augmentative communication tool especially for speech and hearing impaired users. DSL numbers are denoted by making use of hands mostly with different structure, morphology and phonology. Various types of Sign Languages differ due to the difference in their number/gesture sets mainly. Figure 1 represents the invasive DSL numbers used in N-DSLA for the purpose of experimentation.

Table 1. Devnagari numbers.

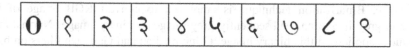

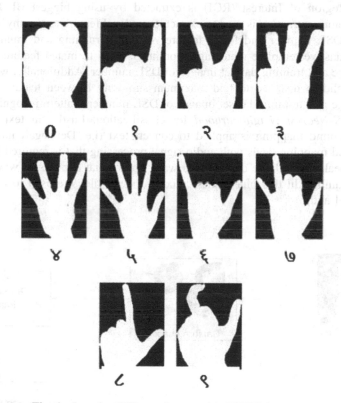

Fig. 1. Invasive DSL numbers used in N-DSLA system

3 Proposed Numeric Devnagari Sign Language Analyzer (N-DSLA) System

3.1 System Architecture for N-DSLA System

The architecture of N-DSLA system is fragmented into six consequent phases namely; image capturing, image pre-processing, region extraction, feature extraction, feature matching and pattern recognition as shown in Fig. 2. Image capturing phase deals with capturing YCbCr image in the form of DSL number image. The frame size of image is considered as 160×120 in complex background along with mixed light using two web cameras (with resolution 720 pixels). Image pre-processing phase consists of skin colour segmentation process, binarization, normalization, noise removal and morphological operations. Skin color segmentation process is applied on YCbCr image. The segmentation process uses skin detector along with YCbCr and produces RGB image. Binarization technique is employed to convert RGB image of DSL number to grey image and subsequently grey image to binary image. Normalization technique is applied on this binary image. Using median and Gaussian filters on binary image of DSL number, the noise is removed in this phase only. Morphological operations such as erosion and dilation are applied on filtered binary image. In region extraction, Region of Interest (ROI) is extracted by using biggest BLOB of size 80 × 60. Feature extraction is used to extract lower-level feature as 'Canny' edge from the ROI of DSL number and forms feature vectors of running and training dataset images. Exhaustive template matching algorithm is used to match feature vectors of running image and training dataset images of DSL number. Additionally, we compute the least Euclidian distance to find maximum similarity between feature vectors of running image and training dataset images of DSL number. Pattern recognition phase consists of *K-Nearest Neighbourhood* for classification. Lastly, in text to speech conversion, sound mapping is applied to convert text (i.e. Devnagari number) into speech. Sound mapping deals with audio signal processing that is required to produce sound via speaker. We have developed a wave library that consists of wave files for Devnagari number. In this phase, these recorded wave files are used to produce the sound (digital audio).

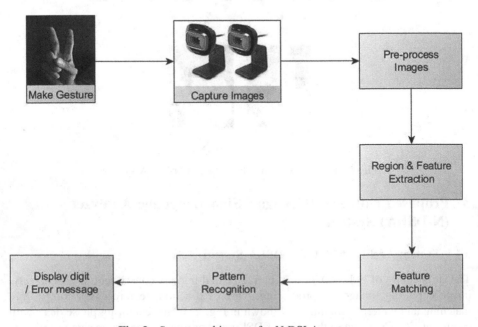

Fig. 2. System architecture for N-DSLA system

3.2   Working of N-DSLA System

Working of N-DSLA system is presented using software architecture for N-DSLA as shown in Fig. 3. It is subdivided in three parts namely; Image processing, Application Programming Interface (API) and Application. API comprises of Aforge.net and EMTU CV. Music player is the application handled by using aforesaid 10 gesture signs (viz. 0 to 9 DSL numbers) of N-DSLA system.

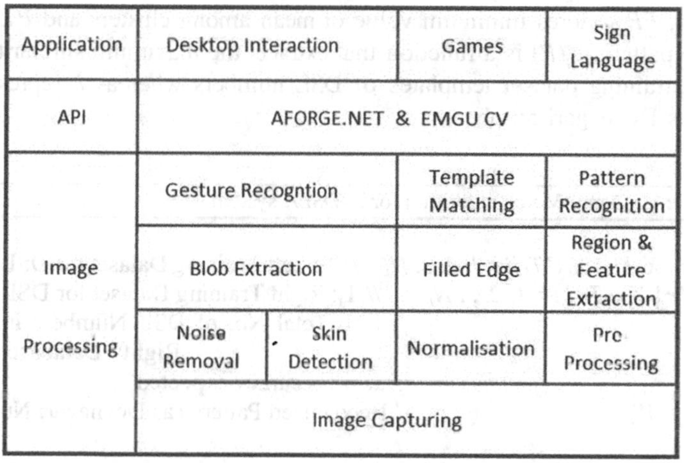

Application	Desktop Interaction		Games	Sign Language
API	AFORGE.NET & EMGU CV			
Image Processing	Gesture Recogntion		Template Matching	Pattern Recognition
	Blob Extraction		Filled Edge	Region & Feature Extraction
	Noise Removal	Skin Detection	Normalisation	Pre Processing
	Image Capturing			

Fig. 3. Software architecture for N-DSLA system

Techniques involved in various phases of N-DSLA system are such as skin color segmentation, binarization, normalization, noise removal, biggest BLOB extraction, feature vector formation, template matching and pattern recognition. Training dataset comprises of 2 datasets namely; LDatabase and RDatabase. LDatabase stores DSL number images captured by using left camera while RDatabase stores DSL number images captured by using right camera. These images are pre-processed and Region of Interest (ROI) of the images of size 120×160 pixels are stored in both databases. Each database has 10 clusters containing 100 images for each DSL number. N-DSLA system focuses on Template Matching techniques. N-DSLA system focuses on *Ex-Temp-Match* algorithm which is discussed as follows:

Ex-Temp-Match Algorithm. This algorithm is based on template descriptor model that measures the similarity between templates of running image and training dataset images of DSL number. The template descriptor T_i is used to evaluate the best suited pattern as Devnagari number as

$$T_i = \min_{l, r \in N} TM(R, (T_l | T_r)) . \omega$$

In this model, $TM(R, (T_l | T_r))$ computes the distance between running image and training dataset (left or right) image templates, N represents the left and right database for DSL numbers, R is a running image template of DSL number, T_l $(i = 1, ..., N)$ and T_r $(i = 1, ..., N)$ denote the left and right training dataset image templates for DSL numbers respectively and ω depicts a template segment for DSL number.

On the basis of above model, we now present Algorithm 1: *Ex-Temp-Match* formally. In this algorithm, X_i, Y_j, T_l and T_r symbolize left and right training datasets for DSL numbers, W_i represents recognized pattern as Devnagari number, $N1$ and $N2$ shows total number of samples of DSL numbers in left and right database respectively, R depicts running image template, μ_i stores resultant of *TM* as appropriate match for DSL number, $PR_{i,p}$ stores minimum value of mean among clusters and P denotes the best suitable pattern, $g_j(R)$ is a function that extracts the maximum similarity between running and training dataset templates of DSL numbers whereas k represents value recognized as Devnagari number.

Algorithm- I: Ex-Temp-Match algorithm for N-DSLA system

Input: $accu$, X_i $\{T_l \mid X_i \in T_l, i= 1, 2 \ldots N\}$ // T_l: Left Training Dataset for DSL Numbers
$\quad\quad\quad$ Y_i $\{Tr \mid Y_i \in Tr, i= 1, 2 \ldots N\}$ // T_r: Right Training Dataset for DSL Numbers
$\quad\quad\quad\quad\quad\quad\quad\quad\quad\quad\quad\quad$ // N: Total No. of DSL Numbers in Left and
$\quad\quad\quad\quad\quad\quad\quad\quad\quad\quad\quad\quad\quad\quad\quad\quad\quad\quad\quad\quad$ Right   Database
$\quad\quad\quad\quad\quad\quad\quad\quad\quad\quad\quad\quad$ //$accu$: accuracy expected
Output: $\quad W_i$ $\quad\quad\quad\quad\quad\quad\quad$ // Recognized Patterns as Devnagari Numbers
Begin
$\quad$ *1.* $\quad$ *for* $i = 1 : N$
$\quad$ *2.* $\quad\quad$ *Begin*
$\quad$ *3.* $\quad\quad\quad$ *get R* $\quad\quad$ // R : Running Image   of DSL Number
$\quad\quad\quad\quad\quad$ $Ti = \min_{l,r \in N} TM(R, (T_l \mid T_r))$ /*TM: Template Matching function
$\quad\quad\quad\quad\quad$ that   matches Running and Training images of left or right database of
$\quad\quad\quad\quad\quad$ DSL Number */

$\quad$ *4.* $\quad$ T= $(T_l \mid T_r)$
$\quad$ *5.* $\quad$ $\mu_i = T_i . \omega$ //μ_i stores resultant of *TM* as appropriate match for DSL   Number
$\quad$ *6.* $\quad$ $PR_{i,p} = find (min (\mu_i))$ // $PR_{i,p}$ stores   minimum value of mean among
$\quad\quad\quad$ clusters
$\quad$ *7.* $\quad\quad$ *for* $k= 1 : P$
$\quad$ *8.* $\quad\quad\quad$ *Begin*
$\quad$ *9.* $\quad\quad\quad\quad$ *find* $PRi_{,k}$ $\quad\quad\quad\quad\quad\quad\quad\quad\quad\quad\quad\quad$ // Pattern Recognition
$\quad$ *10.* $\quad\quad\quad\quad$ *If* $PR_k \geq accu$
$\quad$ *11.* $\quad\quad\quad\quad$ *Begin*
$\quad$ *12.* $\quad\quad\quad\quad\quad$ *get R of PRi*$_{,k}$
$\quad\quad\quad\quad\quad\quad$ $g_j(R) = max (similarity(R, T))$
$\quad$ *13.* $\quad\quad\quad\quad\quad$ $W_i = j$ // j^{th} value recognized as   Devnagari Number
$\quad$ *14.* $\quad\quad\quad\quad$ *end*
$\quad$ *15.* $\quad\quad\quad$ *end*
$\quad$ *16.* $\quad\quad$ *end*
End.

---

4 Experimental Setup of N-DSLA System

The system is developed for detection of DSL numbers based on Single Hand Two Camera approach. This system is highly concerned with the appropriate installation and configuration of two web cameras and capturing images. Configuration of camera deals with the initialization of parameters such as high brightness (145–175 lm), frame rate (30 frames/second) and frame interval (1 or 3) for both cameras. Input running image (i.e. YC_bC_r image) of frame size [160 × 120] pixels is captured by the web camera mounted on top of the monitor. The monitor or the web camera is adjusted at 90° angle with the horizontal plane. The distance of input running image is 15 cm from camera in complex background.

We have used Microsoft *.NET Framework* and *Visual C# 2010* to implement N-DSLA system. The Microsoft *.NET Framework* is a software framework that includes a large library of coded solutions to common programming problems. *Visual C# 2010* is a modern, innovative programming language and tool for building *.NET* based software. The pre-requisite is that, web camera drivers are installed on the computer system. During the execution of N-DSLA system, it is essential for the user to choose the appropriate web cameras for capturing DSL number as input image.

This input image is running image (R) of DSL number whereas left (T_l) and right (T_r) training datasets are constructed with 1000 template images (X_i and Y_i) of DSL numbers. N-DSLA system includes various processes such as image pre-processing, region extraction, feature extraction, feature matching and pattern recognition. These processes are applied on input running image as depicted in Fig. 4. The formation of left and right training dataset is based on aforesaid processes is as represented in Fig. 5.

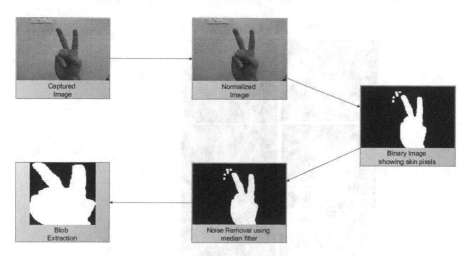

Fig. 4. Working of N-DSLA system

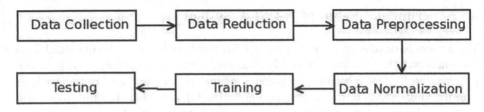

Fig. 5. Training datasets formation for N-DSLA system

Ex-Temp-Match algorithm is applied in N-DSLA system that finds maximum similarity between a running image and the training dataset images. In pattern recognition phase, maximum similarity between running template and training dataset templates is computed to recognize the best suited pattern. This recognized pattern is represented in the form of Devnagari number with picture as shown in Fig. 6 whereas snapshots of N-DSLA are depicted in Fig. 7. Lastly, Devnagari number is translated into speech by using aforementioned sound mapping in Text to Speech conversion. Experimental result of N-DSLA as compared to existing HGRSs is as depicted in Table 5 by considering vital factors such as background, color space, techniques and DSL Numbers.

- **Interaction with a Music Player using Gestures**
  Music player is an application based on the implemented N-DSLA system as represented in Fig. 8. The user performs various operations such as play, pause and stop etc. on the music player using hand gestures. Once the gesture is made, the operation corresponding to the recognized gesture is activated. The speed and accuracy of these operations depend on the training dataset, the resolution of web camera and the clarity of the hand gesture performed by the user.

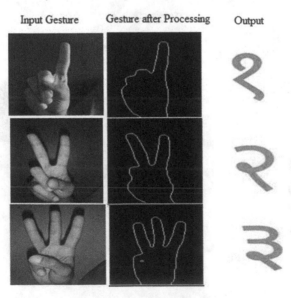

Fig. 6. Detection of DSL numbers of N-DSLA system

(a) Snapshot 1   (b) Snapshot 2   (c) Snapshot 3

Fig. 7. Snapshots of N-DSLA system

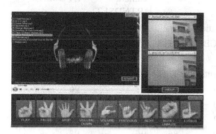

(a) Music player application snapshot 1   (b) Music player application snapshot 2

Fig. 8. Music player application using N-DSLA system

5 Comparative Performance of N-DSLA System

We now present the comparative performance of N-DSLA with other HGRSs. On the basis of some vital factors such as approach, sign languages, techniques, background and recognition rate, a comparative study is presented as shown in Table 3. Following are the important observations associated with N-DSLA system:

- As compared to other HGRSs, N-DSLA system achieves the highest recognition rate i.e. 97.2% using Single Hand Two Camera approach in complex background.
- When fingertip detection and convex hull methods are used in DSL Translator, it attains the recognition rate of 94.13% using Two Hand Single Camera approach in cluttered background.
- Numeric HGRS is designed using histogram and skin color segmentation techniques, accomplishes the detection rate of 93.1% including mixed lighting condition in static background.
- Similarly, α-DSLR is developed using Sim-Temp-Match and K-Cluster-Match techniques, acquires the detection rate of 95.19% using Single Hand Single Camera approach in static background.
- Using Blob and Centroid in Alpha HGRS, recognition rate of 90.19% with using Single Hand Single Camera approach in complex background has been reported.

- Finally, it has been observed that the lowest recognition rate i.e. 87.82% is achieved when histogram matching technique is employed in Real-time HGRS in static background.

However, it is evident from Table 3 that highlighted performance of N-DSLA system shows effective outcome as compared to other HGRSs for identification of DSL numbers (Table 2).

Table 2. Experimental results of N-DSLA system

Recognition rate (%)

Approaches	Single hand single camera (other HGRSs)						Two hands single camera (RTNDSLT)	Single hand two cameras (N-DSLA)
Background	Complex	Complex	Complex	Static	Static	Static	Cluttered	Complex
Color space	RGB	YCbCr	HSV	RGB	RGB	RGB	YCbCr	YCbCr
DSL numbers	Techniques							
	Histogram	DCT	FT	EOH	Centroid	TM	Fingertip	Ex-Temp-Match
0	80	82	90	86	89	86	92	96
1	95	86	90	90	90	90	96	99
2	76	82	84	84	92	88	94	98
3	80	86	90	90	90	86	96	98
4	82	84	92	88	92	90	94	96
5	80	82	88	86	90	90	92	95
6	83	82	88	88	88	86	94	97
7	70	72	86	78	88	84	94	97
8	85	82	90	86	90	88	96	98
9	80	84	92	86	92	90	94	98
10	NA	NA	NA	NA	NA	NA	92	NA
Total	81.1	82.2	89	86.2	90.1	87.8	94.2	97.2

Table 3. Performance comparison of N-DSLA system with existing approaches of HGRSs

HGRS	Approach	Sign language	Techniques	Background	Recognition rate
Alpha HGRS	Single hand single camera	ASL alphabets	Blob, centroid	Complex	90.19%
α-DSLR	Single hand single camera	ASL alphabets	Sim-Temp-Match, K-Cluster-Temp-Match	Complex	95.59%
Real-time HGRS	Single hand single camera	DSL alphabet	Histogram	Static	87.82%
Numeric HGRS	Single hand single camera	DSL numbers	Histogram	Static	93.1%
DSL Translator	Two hand single camera	DSL numbers	Fingertip detection, convex hull	Cluttered	94.13%
N-DSLA	**Single hand two camera**	**DSL numbers**	*Ex-Temp- Match*	**Complex**	**97.2%**

6 Conclusion

Our N-DSLA system is developed for recognition of DSL numbers using Single Hand Two Camera approach. Two web cameras of resolution 720 Pixels capture YC_bC_r images of frame size 160×120 from distance of 15 cm in complex background with mixed light. *Ex-Temp- Match* algorithm is introduced in N-DSLA system so as to achieve the success rate of 97.2% with recognition time of 0.3 s. N-DSLA system is applied for handling Music Player application using 10 gestures from 0 to 9 for corresponding operations.

References

1. Wang, F., Ngo, C., Pong, T.: Simulating a smartboard by real-time gesture detection in lecture videos. IEEE Trans. Multimed. **10**(5), 926–935 (2008). doi:10.1109/TMM.2008. 922871
2. Zhang, X., Chen, X., Member, A., Li, Y., Lantz, V., Wang, K., Yang, J.: A framework for hand gesture recognition based on accelerometer and EMG sensors. IEEE Trans. Syst. Man Cybern Part A **41**(6), 1064–1076 (2011). doi:10.1109/TSMCA.2011.2116004
3. Kim, T.K., Cipolla, R.: Canonical correlation analysis of video volume tensors for action categorization and detection. IEEE Trans. Pattern Anal. Mach. Intell. **31**(8), 1415–1428 (2009). doi:10.1109/TPAMI.2008.167
4. Alon, J., Athitsos, V., et al.: A unified framework for gesture recognition and spatio temporal gesture segmentation. IEEE Trans. Pattern Anal. Mach. Intell. **31**(9), 1685–1699 (2009). doi:10.1109/TPAMI.2008.203
5. Dardas, N.H., Georganas, N.D.: Real-time hand gesture detection and recognition using bag-of-features and support vector machine techniques. IEEE Trans. Instrum. Meas. **60**(11), 3592–3607 (2011). doi:10.1109/TIM.2011.2161140
6. Mohandes, M., Deriche, M., Liu, J.: Image-based and sensor-based approaches to Arabic Sign Language recognition. IEEE Trans. Hum. Mach. Syst. **44**(4), 551–557 (2014). doi:10. 1109/THMS.2014.2318280
7. Kosmidou, V., Petrantonakis, P., Hadjileontiadis, L.: Weighted intrinsic-mode entropy and signer's level of deafness. IEEE Trans. Syst. Man Cybern. Part B Cybern. **41**(6), 1531–1543 (2011)
8. Yang, H., Sclaroff, S., Member, S.: Sign language spotting with a threshold model based on conditional random fields. IEEE Trans. Pattern Anal. Mach. Intell. **31**(7), 1264–1277 (2009). doi:10.1109/TPAMI.2008.172
9. Zhu, C., Sheng, W.: Wearable sensor-based hand gesture and daily activity recognition for robot-assisted living. IEEE Trans. Syst. Man Cybern. Part A Syst. Hum. **41**(3), 569–573 (2011). doi:10.1109/TSMCA.2010.2093883
10. Tindale, A., Kapur, A., Tzanetakis, G.: Training surrogate sensors in musical gesture acquisition systems. IEEE Trans. Multimed. **13**(1), 50–59 (2011). doi:10.1109/TMM.2010. 2089786
11. Reale, M., Canavan, S., Yin, L., Hu, K., Hung, T.: A multi-gesture interaction system using a 3-D iris disk model for gaze estimation and an active appearance model for 3-D hand pointing. IEEE Trans. Multimed. **13**(3), 474–486 (2011). doi:10.1109/TMM.2011.2120600
12. Tran, C., Trivedi, M.M.: 3-D posture and gesture recognition for interactivity in smart spaces. IEEE Trans. Ind. Inform. **8**(1), 178–187 (2012). doi:10.1109/TII.2011.2172450

13. Yanik, P., Manganelli, J., Merino, J., Threatt, A., Brooks, J.O., Green, K., Walker, I.: A gesture learning interface for simulated robot path shaping with a human teacher. IEEE Trans. Hum. Mach. Syst. **44**(1), 41–54 (2014). doi:10.1109/TSMC.2013.2291714
14. Wu, C., Chiu, Y., Guo, C.: Text generation from Taiwanese Sign Language using a PST-based language model for augmentative communication. IEEE Trans. Neural Syst. Rehabil. Eng. **12**(4), 441–454 (2004). doi:10.1109/TNSRE.2003.819930
15. Wang, L., Wang, R., Kong, D., Yin, B.: Similarity assessment model for Chinese Sign Language videos. IEEE Trans. Multimed. **16**(3), 751–761 (2014). doi:10.1109/TMM.2014. 2298382
16. Li, Y., Chen, X., Zhang, X., Wang, K., Wang, Z.J.: A sign-component-based framework for Chinese Sign Language recognition using accelerometer and sEMG data. IEEE Trans. Biomed. Eng. **59**(10), 2695–2704 (2012). doi:10.1109/TBME.2012.2190734
17. Zaki, M., Shaheen, S.: Sign language recognition using a combination of new vision based features. Pattern Recogn. Lett. **32**(4), 572–577 (2011). doi:10.1016/j.patrec.2010.11.013
18. Chen, F., Fu, C., Huang, C.: Hand gesture recognition using a real-time tracking method and hidden Markov models. Image Vis. Comput. **21**(8), 745–758 (2003). doi:10.1016/S0262-8856(03)00070-2
19. Al-Jarrah, O., Shatnawi, A., Halawani, A.: Recognition of gestures in Arabic Sign Language using neural networks. Artif. Intell. **133**(1–2), 131–136 (2006). doi:10.1016/S0004-3702(01) 00141-2
20. Martins, P., Rodrigues, H., Rocha, T., Francisco, M., Morgado, L.: Accessible options for deaf people in e-learning platforms: technology solutions for sign language translation. Procedia Comput. Sci. **67**, 263–272 (2015). doi:10.1016/j.procs.2015.09.270
21. Pansare, J.R., Ingle, M., Gawande, S.: Real-time static hand gesture recognition for American Sign Language (ASL) in complex background. J. Signal Inf. Process. **3**, 364–367 (2012). doi:10.4236/jsip.2012.33047
22. Mitra, S., Acharya, T.: Gesture recognition : a survey. IEEE Trans. Syst. Man Cybern. Part C Appl. Rev. **37**(3), 311–324 (2007). doi:10.1109/TSMCC.2007.893280

Study of Engineered Features and Learning Features in Machine Learning - A Case Study in Document Classification

Arpan Sen[(✉)], Shrestha Ghosh, Debottam Kundu, Debleena Sarkar,
and Jaya Sil

Department of Computer Science and Technology,
Indian Institute of Engineering Science and Technology,
Shibpur, Howrah 711103, West Bengal, India
s.arpan1993@gmail.com, ghosh_shrestha@yahoo.co.in,
debottam.kundu330@gmail.com, debleena0412@gmail.com, js@cs.iiests.ac.in

Abstract. Document classification is challenging due to handling of voluminous and highly non-linear data, generated exponentially in the era of digitization. Proper representation of documents increases efficiency and performance of classification, ultimate goal of retrieving information from large corpus. Deep neural network models learn features for document classification unlike the engineered feature based approaches where features are extracted or selected from the data. In the paper we investigate performance of different classifiers based on the features obtained using two approaches. We apply deep autoencoder for learning features while engineering features are extracted by exploiting semantic association within the terms of the documents. Experimentally it has been observed that learning feature based classification always perform better than the proposed engineering feature based classifiers.

Keywords: Deep learning · Feature extraction · Autoencoder · Restricted Boltzmann Machine · Semantic association · N-gram Model

1 Introduction

In this era of information technology, the necessity of analyzing the large data repositories is felt for extracting knowledge which has been either embedded or hidden in the data. Exponential growth of documents throw challenges to the scientific community to classify the huge documents based on certain topics, which efficiently can retrieve information [17]. In addition, document classification has an immense scope of application in document indexing, maintaining a hierarchical form of web content, spam filtering, semantic analysis and many more. However, engineering feature based approaches is highly dependent on feature identification, which by and large depends on humans. Moreover, handling of large documents is difficult for feature generation and time

© Springer International Publishing AG 2017
A. Basu et al. (Eds.): IHCI 2016, LNCS 10127, pp. 161–172, 2017.
DOI: 10.1007/978-3-319-52503-7_13

consuming too. An efficient document representation aims at transforming the raw input into discriminative features of each class by learning the characteristics of the documents.

The simplest way of representing a text document is to consider the unique words appearing in the documents after preprocessing. An input vector for each document is built using binary symbols where presence of a word in the document is represented by '1', otherwise '0'. Another representation is the bag of words (BoW) [12, 25] where the frequency of occurrence of each word in the documents is used as an input vector. The application of BoW is limited because this representation cannot measure the relative importance of a word in a document. The Term Frequency-Inverse Document Frequency (TF-IDF) [21] based representation overcomes this deficiency. However, these kind of representations merely deal with the presence of words in a document without exploiting the correlation among the features and have redundancy.

Learning feature based approaches are recently gaining importance for automated text processing which includes steps like representation, classifier construction and evaluation [24]. Deep learning processes this raw data by separating underlying features and compressing the raw data by extracting relevant information. The raw input is fed to the input layer of a deep network, which on passing through the hidden layers give a new set of discriminative features.

Deep learning models have been used as generative models of many different types of data including labeled or unlabeled images [9], windows of mel-cepstral coefficients representing speech [19], bags of words representing documents [11], and user ratings of movies [23]. Restricted Boltzmann machines (RBMs), a kind of deep learning models [5] has been employed in our work for dimensionality reduction and feature extraction before classification using deep neural network classifiers [7].

In this paper, we focus on efficient document representation techniques using machine learning approaches. Owing to the huge bulk of the text in the documents, it is necessary to transform the documents into some discriminative features for each class which can be fed into the classifiers for better organisation of the documents. Until recently, machine learning mechanisms essentially used one or two layers of linear or nonlinear feature transformations. Such mechanisms while effective in solving simple well-constrained problems, have shortcomings when the training data is highly non-linear with pattern variation due to limited representational power. This problem is efficiently dealt using deep learning, a class of machine learning techniques that exploit hierarchically nonlinear information processing based on supervised or unsupervised methods.

Before deep learning techniques gained popularity, feature engineering played a major role in feature extraction. Statistical methods are used to capture word relationships in a document. Semantic representation of a text document has been devised in the paper by proposing a novel method that measures the contribution of each word to the meaning of the document. Performance of the approaches for document classification are compared in the paper and the experimental results are shown.

The main contributions of this paper are:

- Generating document representative features by deep learning and by semantic models.
- Proposing a semantic association model that is an extension of the N-gram model.
- Experimental analysis shows that the representative features are better than the traditional TF-IDF scores.
- Experimental results demonstrate that the features obtained when a Deep Belief Network(DBN) is trained in a greedy layerwise manner followed by fine tuning of weights using a deep auto encoder (AE) [10] is better than the semantic association model.

We have organised this paper as follows. In Sect. 2, we explain learning feature based models and in Sect. 3 the proposed semantic association model, Non-repeating neighbour (NRN) based approach has been described to represent the text documents. Experimental analysis of the features obtained from the deep learning model and the NRN model is presented in Sect. 4. We conclude this paper in Sect. 5.

2   Learning Feature Based Model

Feature learning in document classification highly depends on the document representation for transforming raw input into feature vectors. Handling raw input is difficult due to its high dimensionality and so by representation, underlying variations and correlations of features are captured. In machine learning algorithms a large vocabulary is introduced during training. Such a high dimensional input vector is sparse if represented using traditional techniques. Good representations can express many general priors which may be used by the learning algorithms [1]. Some of the priors are smoothness, hierarchy of features, sparsity, which determines the small fraction of relevant features out of all possible features. Simplicity of the features increases as the number of levels in a representation increases. Therefore, the better the representation, the more accurate is the performance of the classifier.

2.1   Methodology

In this section, we describe the deep architecture for learning of features using unsupervised learning strategy and produces a feature vector of dimensionality lower than that of the input vector. A Deep Belief Network(DBN) is a probabilistic generative model composed of multiple layers of stochastic, latent variables [8]. It can be viewed as a composition of simple learning modules, each of which is a RBM containing a layer of visible units representing the data. A layer of hidden units tries to capture the higher-order co-relations in the data.

A deep Autoencoder(AE) is a Deep Neural Network(DNN) which does not use class labels. It consists of an input layer to which the input vector is fed,

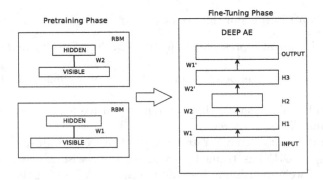

Fig. 1. Pre-training and fine-tuning phase of DBN - Deep AE

more than one hidden layers which represent the transformed features and an output layer, of size equal to that of the input layer, which is used for input reconstruction. The weight matrices and bias vectors are randomly initialized and Sigmoidal activation function is used in the hidden and the output layer of the deep AE. Training of deep AE using gradient descent method with back propagation (BP)[22] algorithm becomes ineffective with the increase of number of hidden layers. By the time the reconstruction error calculated at the output layer is propagated back to the initial layers, the gradient becomes negligible due to the vanishing gradient problem [2] and hence reduces the training efficiency.

DBN, Deep AE are randomly initialized and often fails to obtain a good solution because of its non-proximity to a good solution. An effective way to initialize weights such that they are close to a good solution was proposed in [10] and is used here. A DBN is initially trained in an unsupervised way to obtain the final weights of the network. This phase is called the pretraining phase. The final weights of the DBN are used as initial weights of this deep AE which are then fine tuned using gradient descent algorithm with BP, demonstrated in Fig. 1.

3   Semantic Association Model

Feature engineering approach for generating features is intensive and time consuming. Moreover, it fails to capture the underlying feature variations and the context of a document. The limitations of engineering feature based approaches has been addressed in the paper by semantic modelling which represents the distribution of words in a document and captures the word relationships. Semantics of a document is derived by measuring the effect of each word and its surroundings in order to encode the meaning of the document.

3.1   Non-Repeating Neighbour (NRN) Method

We propose a semantic association model called Non-Repeating Neighbour (NRN) method for generating features to represent the documents.

The NRN model is inspired by the N-gram model [15]. It is an extension of the N-gram model where unlike the N-gram model, the value of N is not fixed apriori. A unigram dictionary is formed using the training set following which an initial word score is computed for every unique word present in the training corpus. The initial word score G_x for a word x is given in Eq. 1.

$$G_x = I_x \times IDF_x \tag{1}$$

where, I_x is the self information conveyed by the word x, calculated using Eq. 2 [16] and IDF_x is the inverse document frequency [13] of x calculated using Eq. 3

$$I_x = P_x \times log_{10}(\frac{1}{P_x}) \tag{2}$$

$$IDF_x = log_{10}(\frac{N}{1 + n_x}) \tag{3}$$

P_x is the probability of occurrence of word x by considering the given training corpus and calculated using Eq. 4, N is the total number of documents in the corpus and n_x represents the number of documents in which the word x occurs.

$$P_x = \frac{F_x}{T} \tag{4}$$

where, F_x is the frequency of occurrence of word x in the corpus and T is the total number of words in the corpus.

In order to calculate the local score of a word we make two lists; one containing the left neighbors and another the right neighbors of the word. The neighbors are appended to each of these list by propagating in both the directions. Addition of neighbor stops when there is a repetition in the list or a sentence boundary is reached. Based on the score of these neighbors and the word itself, a local score is generated (refer Algorithm 1). As the distance of the neighboring words from the word increases gradually, the effect of the neighbors decreases. This effect is calculated by multiplying the neighbors scores with a factor equal to the inverse of the distance between the word and its neighbor.

This is different from other models like tf-idf where instead of giving importance to the position of words in the document importance is given to the frequency of the words in the documents. In this approach we try to preserve some of the contextual meaning of the words by considering its neighborhood.

The NRN model is illustrated by the following example.

Let us consider a simple dataset consisting of the following three single sentence documents.

- *Document1*: This is the house that Jack Built
- *Document2*: The brown mouse played in the house that Jack built
- *Document3*: The little brown mouse ran inside the little house

Algorithm 1. Local Score Generation Algorithm

Input: globe_score: a list of global scores of each unique word in the corpus,
train_docs: a set of training documents, list_words: a list of unique words in the
corpus
Output: local_score: is a matrix of local scores of each unique word in each doc
Set the initial local_score matrix to zero

```
01. for every document i
02.    for every word j in list_words
03.        if list_words[j] ∉ train_docs[i]
04.            local_score[i][j] ← 0
05.            break
06.        else
07.            rand_word ← rand_select(list_words[j] from train_docs[i])
08.        endif
09.        r ← 0; l ← 0; sum ← globe_score[rand_word]
10.        while (right_list has unique words)
11.            right_list[r] ← right(r, rand_word)
12.            r + +          { Add new word found on right to right_list }
13.        end while
14.        while (left_list has unique words)
15.            left_list[r] ← left(l, rand_word)
16.            l + +          { Add new word found on left to left_list }
17.        end while
18.        for every word k in right_list
19.            sum ← sum + globe_score[right_list[k]] × inverse_depth_of(k)
20.        end for
21.        for every word k in left_list
22.            sum ← sum + globe_score[left_list[k]] × inverse_depth_of(k)
23.        end for
24.        local_score[i][j] ← sum
25.    end for
26. end for
```

The list of unique words in the dataset is

this, is, the, house, that, jack, built, brown, mouse, played, in, little, ran, inside

For each unique word the initial word score is calculated and the traversal of the word around *house* in document 2 forms a tree-like structure as shown in Fig. 2.

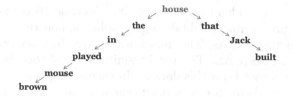

Fig. 2. Tree with root node = *house*

4   Experimental Analysis

In this section we introduce the dataset we have worked on and describe the experimental setup. Thereafter, we provide a detailed analysis of all the results.

4.1   Dataset

In the experiment, we use the standard WebKB dataset [3] consisting of four different classes - *student, faculty, course* and *project*. The dataset is randomly split into train and test data with 67% of the total documents being used for training and the remaining 33% being used for testing. After removing stopwords and stemming [20], the number of unique words in the vocabulary is 7287. The distribution of the documents in various classes is shown in Table 1.

Table 1. Distribution of Documents in the WebKB Dataset

WebKB			
Class	# Train Docs	# Test Docs	Total # Docs
Student	1097	544	1641
Faculty	750	374	1124
Course	620	310	930
Project	336	168	504
Total	**2803**	**1396**	**4199**

4.2   Experimental Setup

The input vector for each document consisting of 7287 unique words. The dimension of the input vector is 7287 where each element represents presence or absence of the corresponding word by '1' or '0' respectively. All models have been simulated in MATLAB. In case of learning feature based approach, we use two RBMs to form a DBN. The first RBM has 7287 visible units and 1000 hidden units and the second one with 1000 visible units and 500 hidden units. The deep AE is of $7287 - 1000 - 500 - 1000 - 7287$ architecture having input layer and output layer of size 7287 and three hidden layers of sizes 1000, 500 and 1000 respectively. The DBN is used for pretraining to obtain a good solution and the deep AE is used for fine tuning of the weights. The final features are obtained from the second hidden layer of the deep AE. The total training time of the above mentioned setup is approximately 6 h for this dataset. In case of engineering feature based approach, the input vector for each document is of size 7287 corresponding to 7287 unique words, where each entry stores the corresponding word's local score for the NRN Model.

4.3   Results and Discussions

In Table 2 we provide comparisons of the deep learning method and the NRN method with traditional feature representation methods - the tf-idf and the bag-of-words. We have used three classifiers which are Naive Bayes, kNN and SVM. For SVM classification we have used a ONE vs ALL approach using gaussian kernel. The classification was performed using the WEKA tool [6]. It has been observed that for each of the cases and for each of the methods the performance of the Naive Bayes classifier are comparable. This is because of the inherent assumption of Naive Bayes which considers the input features to be independent, given a class variable. For the other two classifiers, kNN and SVM (Gaussian kernel), the features extracted by us clearly outperforms the other representations which shows the superiority of this feature extraction technique over traditional methods.

We also evaluate NRN method using random sampling where 2803×7287 matrix is sampled to form a $2803 \ S$ matrix where S is the number of sampled features, 2803 is the number of documents and 7287 is the number of unique words. Ten random $2803 \times S$ samples are collected (with repetition) and average classification accuracy is calculated to evaluate the performance of the NRN method. The number of features thus obtained from both deep learning and semantic association are compared in terms of classification accuracy for different values of S. The results are recorded in Table 3. Variability of the classification accuracy with number of samples is shown in Fig. 3, where SVM is considered as the best performing classifier.

Table 2. Comparison of extracted features with traditional representation models

Classification method	Classification accuracy			
	DBN-deep AE	NRN model	Bag-of-words	TF-IDF
kNN(k = 10)	84.24%	75.16%	48.50%	39.47%
Naive Bayes	79.73%	76.72%	77.79%	75.57%
SVM (Gaussian kernel)	88.61%	86.12%	80.01%	49.28%

Table 3. Classification accuracy values

Classifier	Classification accuracy for features											
	400		750		1100		1500		1900		2250	
	I	II	I	II	I	II	I	II	I	II	I	II
	NRN	DL	NRN	DL	NRN	DL	NRN	DL	NRN	DL	NRN	DL
kNN (k = 10)	50%	78%	54%	83%	54%	84%	58%	85%	62%	85%	63%	86%
Naive Bayes	40%	45%	54%	55%	55%	68%	59%	74%	61%	55%	64%	75%
SVM (Linear kernel)	54%	85%	66%	88%	75%	90%	76%	90%	79%	90%	79%	90%

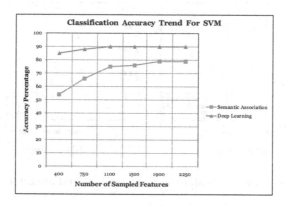

Fig. 3. Classification accuracy trend for SVM

In the Fig. 4, we provide the Receiver Operating Characteristic (ROC) curves for the four different classes - *student, faculty, course* and *project* for the SVM classifier. The accuracy of each class can be estimated from the Area Under the ROC curve. It is evident that performance of class *project* is not as good as the other classes. This is due to the fact that number of training samples available for *project* class is considerably lower compared to others and also because *project* class can be thought as an overlap of some contents from *student, faculty* and *course* classes.

The experimental results obtained from semantic association and deep learning are both better than those obtained from the BoW and TF-IDF based methods. Performance of deep learning based approach is better than the semantic association based approach. Deep neural network classifiers outperform the SVM due to its ability to learn the semantic relationships on its own.

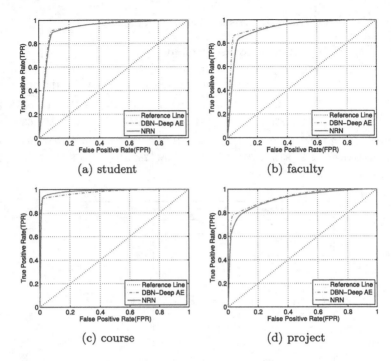

(a) student   (b) faculty

(c) course   (d) project

Fig. 4. ROC curves of the classes *student, faculty, course* and *project* for SVM classifiers

5   Conclusion and Future Work

In the paper we emphasize the importance of feature representation for classification. The potential of deep learning in feature extraction process for efficient compression and representation of raw features is explored. By conducting multiple experiments we deduce that a DBN - Deep AE feature extractor and a DNNC outperforms most other techniques providing a trade-off between accuracy and execution time. In this paper we have dealt with the most significant feature extraction and classification techniques for text documents where each text document belongs to a single class label. With the explosion of digital information a large number of documents may belong to multiple class labels handling of which is a new challenge and scope of future work. Word2vec models [18] in association with Recurrent Neural Networks(RNN) [4, 14] have recently started gaining popularity in feature representation domain. We would like to compare their performance with our deep learning method in future. Similar feature extraction techniques can also be applied to image data to generate compressed feature which can facilitate efficient classification. We would also like to explore such possibilities in our future work.

References

1. Bengio, Y., Courville, A., Vincent, P.: Representation learning: a review and new perspectives. IEEE Trans. Pattern Anal. Mach. Intell. **35**(8), 1798–1828 (2013)
2. Bengio, Y., Simard, P., Frasconi, P.: Learning long-term dependencies with gradient descent is difficult. IEEE Trans. Neural Networks **5**(2), 157–166 (1994)
3. Cachopo, A.: Improving methods for single-label text categorization. Ph.D. thesis, Universidade Tecnica de Lisboa (2007)
4. Elman, J.L.: Finding structure in time. Cogn. Sci. **14**(2), 179–211 (1990)
5. Gehler, P.V., Holub, A.D., Welling, M.: The rate adapting poisson model for information retrieval and object recognition. In: Proceedings of the 23rd International Conference on Machine Learning, pp. 337–344. ACM (2006)
6. Hall, M., Frank, E., Holmes, G., Pfahringer, B., Reutemann, P., Witten, I.H.: The weka data mining software: an update. SIGKDD Explor. Newsl. **11**(1), 10–18 (2009)
7. Hinton, G.E.: To recognize shapes, first learn to generate images. Prog. Brain Res. **165**, 535–547 (2007)
8. Hinton, G.E.: Deep belief networks. Scholarpedia **4**(5), 5947 (2009)
9. Hinton, G.E., Osindero, S., Teh, Y.W.: A fast learning algorithm for deep belief nets. Neural Comput. **18**(7), 1527–1554 (2006)
10. Hinton, G.E., Salakhutdinov, R.R.: Reducing the dimensionality of data with neural networks. Science **313**(5786), 504–507 (2006)
11. Hinton, G.E., Salakhutdinov, R.R.: Replicated softmax: an undirected topic model. In: Advances in Neural Information Processing Systems, pp. 1607–1614 (2009)
12. Joachims, T.: Learning to Classify Text Using Support Vector Machines: Methods. Kluwer Academic Publishers, Theory and Algorithms (2002)
13. Jones, K.S.: A statistical interpretation of term specificity and its application in retrieval. J. Documentation **28**, 11–21 (1972)
14. Jordan, M.I.: Serial order: a parallel distributed processing approach. Adv. Psychol. **121**, 471–495 (1997)
15. Jurafsky, D.: Speech & Language Processing. Pearson Education, India (2000)
16. Meilă, M.: Comparing clusterings-an information based distance. J. Multivar. Anal. **98**(5), 873–895 (2007)
17. Messerly, J.J., Heidorn, G.E., Richardson, S.D., Dolan, W.B., Jensen, K.: Information retrieval utilizing semantic representation of text, 13. US Patent 6,076,051., June 2000
18. Mikolov, T., Sutskever, I., Chen, K., Corrado, G.S., Dean, J.: Distributed representations of words and phrases and their compositionality. In: Advances in neural information processing systems, pp. 3111–3119 (2013)
19. Mohamed, A.R., Sainath, T.N., Dahl, G., Ramabhadran, B., Hinton, G.E., Picheny, M., et al.: Deep belief networks using discriminative features for phone recognition. In: 2011 IEEE International Conference on Acoustics, Speech and Signal Processing (ICASSP), pp. 5060–5063. IEEE (2011)
20. Porter, M.F.: An algorithm for suffix stripping. Program **14**(3), 130–137 (1980)
21. Ramos, J.: Using TF-IDF to determine word relevance in document queries. In: Proceedings of the First Instructional Conference On Machine Learning (2003)
22. Rojas, R.: Neural Networks: A Systematic Introduction. Springer, Heidelberg (2013)

23. Salakhutdinov, R., Mnih, A., Hinton, G.: Restricted boltzmann machines for collaborative filtering. In: Proceedings of the 24th International Conference on Machine learning, pp. 791–798. ACM (2007)
24. Sebastiani, F.: Machine learning in automated text categorization. ACM Comput. Surv. (CSUR) 34(1), 1–47 (2002)
25. Zhang, Y., Jin, R., Zhou, Z.H.: Understanding bag-of-words model: a statistical framework. Int. J. Mach. Learn. Cybernet. 1(1–4), 43–52 (2010)

Implementation of a Digital Hearing Aid with User-Settable Frequency Response and Sliding-Band Dynamic Range Compression as a Smartphone App

Saketh Sharma, Nitya Tiwari, and Prem C. Pandey[✉]

Department of Electrical Engineering, Indian Institute of Technology Bombay,
Mumbai, India
{sakethsharma,nitya,pcpandey}@ee.iitb.ac.in

Abstract. Persons with sensorineural hearing loss suffer from degraded speech perception caused by frequency-dependent elevation of hearing thresholds, reduced dynamic range and abnormal loudness growth, and increased temporal and spectral masking. For improving speech perception by persons with moderate loss of this type, a digital hearing aid is implemented as an Android-based smartphone app. It uses sliding-band dynamic range compression for restoring normal loudness of low-level sounds without making the high-level sounds uncomfortably loud and for reducing the perceptible temporal and spectral distortions associated with currently used single and multiband compression techniques. The processing involves application of a frequency dependent gain function calculated on the basis of critical bandwidth based short-time power spectrum and is realized using FFT-based analysis-synthesis. The implementation has a touch-controlled graphical user interface enabling the app user to fine tune the frequency-dependent parameters in an interactive and real-time mode.

Keywords: Dynamic range compression · Hearing aid · Interactive settings · Sensorineural hearing loss · Smartphone app

1 Introduction

Abnormalities in the sensory hair cells or the auditory nerve result in sensorineural hearing loss. It is a commonly occurring hearing loss, which may be inherited genetically or may be caused by excessive noise exposure, aging, infection, or use of ototoxic drugs. It is associated with frequency-dependent elevation of hearing thresholds, reduced dynamic range along with abnormal growth of loudness known as loudness recruitment, and increased temporal and spectral masking leading to degraded speech perception [1–3]. Hearing aids used for alleviating the effects of sensorineural hearing loss provide frequency-selective amplification along with dynamic range compression [4–8]. They may have directional microphones for improving the signal-to-noise ratio and may employ signal processing for suppression of wind and background noise, acoustic feedback cancellation, environment learning for selection of the frequency response, and generation of comfort noise for patients with tinnitus. Some aids also provide facility of customization using mobile apps and wireless connectivity.

© Springer International Publishing AG 2017
A. Basu et al. (Eds.): IHCI 2016, LNCS 10127, pp. 173–186, 2017.
DOI: 10.1007/978-3-319-52503-7_14

Frequency-dependent elevation of hearing thresholds occurs due to loss of inner hair cells and leads to reduced audibility of speech. Loss of outer hair cells results in reduced dynamic range, i.e. increase in hearing threshold levels without corresponding increase in upper comfortable levels. It also distorts the loudness relationships among speech components leading to degraded speech perception [2, 3]. As linear amplification of sounds is unsuitable for persons with sensorineural hearing loss, hearing aids provide amplification along with dynamic range compression to compensate for reduced dynamic range. The gains and compressions are usually set by an audiologist based on the patient's audiogram and in accordance with a hearing aid fitting protocol [4, 6, 7].

Dynamic range compression involves changing the gain based on the input signal level for presenting the sounds comfortably within the limited dynamic range of the hearing impaired listener [4, 5, 8–11]. The processing is carried out as single-band compression or as multiband compression. In single band compression, the gain is calculated as a function of the signal power over its entire bandwidth. As speech signal is dominated by low frequency components, this compression distorts the perceived temporal envelope of the signal and often makes the high frequency components inaudible. Most of the currently available hearing aids provide multiband dynamic range compression which involves dividing the input signal in multiple bands and applying a gain in each band which is a function of the signal power in that band. It avoids the problems associated with single band compression, but results in decreased spectral contrasts and modulation depths in the speech signal. Further, different gains in adjacent bands may distort spectral shape of a formant spanning the band boundaries, particularly during formant transitions. With the objective of reducing the temporal and spectral distortions associated with the compression techniques currently used in the hearing aids, a sliding-band compression technique has been reported by Tiwari and Pandey [12, 13] for use in digital hearing aids. It calculates a frequency-dependent gain function, wherein the gain for each frequency sample is calculated as a function of the signal power in the auditory critical band centered at it. It keeps compression related distortions below perceptible levels for improving speech perception.

Hearing aids are designed using ASICs due to power and size constraints. Therefore, incorporation of a new compression technique in hearing aids and its field evaluation is prohibitively expensive. Use of smartphone-based application software (app) to customize and remotely configure settings on hearing aids provide a greater flexibility to hearing aid users and developers. Many hearing aid manufacturers (GN ReSound, Phonak, Unitron, Siemens, etc.) provide apps to control hearing aids using Android or iOS smartphone. This type of app helps the hearing aid user in personalizing the listening experience by adjustment of settings during use of the device and avoids repeated visits to an audiology clinic. The smartphone-based apps may also be used for development and testing of signal processing techniques for hearing aids. Hearing aid apps [15–20] such as 'Petralex', 'uSound', 'Q+', and 'BioAid' for Android/iOS, 'Mimi', 'Enhanced Ears' for iOS, and "Hearing Aid with Replay" and "Ear Assist" for Android provide users with moderate sensorineural hearing loss a low-cost alternative for hearing aids. In addition to providing frequency-selective gain and multiband dynamic range compression, they also offer the flexibility of creating and storing sound profiles specific to the user's hearing loss characteristics.

Implementation of the sliding-band compression technique for real-time processing using a 16-bit fixed-point DSP chip with on-chip FFT hardware has been reported in [12] and its implementation as an iPhone-based hearing aid app has been reported in [14]. However, these implementations use predefined processing parameters. To make this low-distortion compression technique conveniently available to the hearing impaired persons and to permit its evaluation by a large number of users without incurring the expenses involved in the ASIC-based hearing aid development, a smartphone app implementing the digital hearing aid with user settable frequency response and dynamic range compression is needed. Such an app is developed for providing the user flexibility to fine tune the frequency-dependent parameters in an interactive and real-time mode by a touch-controlled graphical user interface (GUI). The implementation is carried out using an Android-based smartphone handset, as such handsets are available from a number of manufacturers with features and cost suiting a large number of users. The handset model 'Nexus 5X' with Android 7.0 is used for implementation and testing on the basis of its relatively low audio I/O (input-output) delay and high processing capacity. Most handset models may meet these criteria in the near future.

The signal processing for the app is described in the second section. The third section presents the details of its implementation as a smartphone app. Test results are presented in the fourth section, followed by conclusions in the last section.

2   Signal Processing for Settable Frequency Response and Sliding-Band Dynamic Range Compression

Block diagram of the signal processing for the app is shown in Fig. 1. The operations for compensation of increased hearing thresholds and decreased dynamic range are carried out by spectral modification. The processing uses a DFT-based analysis-synthesis comprising the steps of short-time spectral analysis, frequency and level dependent spectral modification, and signal resynthesis. The analysis involves segmentation of the input signal into overlapping frames and calculating DFT to get short-time complex spectra. A frequency-dependent gain function is calculated in accordance with the control settings and as a function of the short-time magnitude spectrum. This gain function is used for modification of the short-time input complex spectrum and its IDFT is calculated to get the overlapping output signal frames. The output signal is re-synthesized using an overlap-add technique for reducing the processing related artifacts.

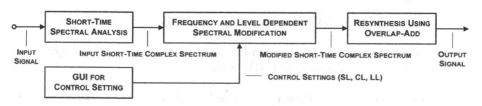

Fig. 1. Signal processing in the hearing aid app with settable frequency response and dynamic range compression.

In our design, it is possible to use a look-up table to provide the gain function most appropriate for improving speech perception by the listener. However, assessment of the loudness growth function as a function of frequency to derive the look-up table is not practical. Hence, the control settings are in the form of plots of desired levels for 'soft', 'comfortable', and 'loud' sounds (referred to as SL, CL, LL, respectively) and input using touch-controlled GUI in an interactive manner. These values are used for calculating the frequency and level dependent gain function. For each frequency sample k, the spectral modification is carried out using a piece-wise linear relation between the input power and the output power on dB scale as shown in Fig. 2. The relationship is specified by the values of $P_{OdBSL}(k)$, $P_{OdBCL}(k)$, and $P_{OdBLL}(k)$ which are the output signal levels corresponding to soft, comfortable, and loud sounds, respectively, for the hearing aid user and by the values of $P_{IdBSL}(k)$ and $P_{IdBLL}(k)$ which are the input signal levels corresponding to soft and loud sounds, respectively, for a normal-hearing listener.

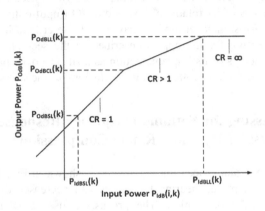

Fig. 2. Relation between input power (dB) and output power (dB) for ith frame and band centered at kth spectral sample.

The lower segment of the input-output relationship, marked as 'CR = 1' in Fig. 2, corresponds to linear gain providing the amplification needed for soft level sounds and the value of the gain is given as

$$G_{LdB}(k) = P_{OdBSL}(k) - P_{IdBSL}(k) \tag{1}$$

This gain is applied unless the output exceeds the comfortable level. Thus the input-output relationship in this segment for ith frame is given as the following:

$$P_{OdB}(i,k) = P_{IdB}(i,k) + G_{LdB}(k), \quad P_{IdB}(i,k) < P_{OdBCL}(k) - G_{LdB}(k) \tag{2}$$

The central segment of the relationship, marked as 'CR > 1', corresponds to compression with compression ratio $CR(k)$ given as

$$CR(k) = \{P_{IdBLL}(k) - P_{OdBCL}(k) + G_{LdB}^{\cdot}(k)\} / \{P_{OdBLL}(k) - P_{OdBCL}(k)\} \tag{3}$$

This relationship is applicable for the output lying between comfortable and loud levels, and is given as the following:

$$P_{OdB}(i,k) = P_{OdBCL}(k) + \{P_{IdB}(i,k) - P_{OdBCL}(k) + G_{LdB}(k)\}/CR(k),$$
$$P_{OdBCL}(k) - G_{LdB}(k) \leq P_{IdB}(i,k) \leq P_{IdBLL}(k) \tag{4}$$

The upper segment of the relationship, marked as 'CR = ∞', corresponds to limiting of the output power and is applied when the output exceeds loud level. The relationship is given as the following:

$$P_{OdB}(i,k) = P_{OdBLL}(k), \quad P_{IdB}(i,k) > P_{IdBLL}(k) \tag{5}$$

In accordance with the input-output relationship as represented by the three segments and given in Eqs. 2, 4, and 5, the target gain for the spectral sample k in ith frame is given as:

$$G_{TdB}(i,k) = \begin{cases} G_{LdB}(k), & P_{IdB}(i,k) < P_{OdBCL}(k) - G_{LdB}(k) \\ \frac{G_{LdB}(k) - \{P_{IdB}(i,k) - P_{OdBCL}(k)\}\{CR(k)-1\}}{CR(k)}, & P_{OdBCL}(k) - G_{LdB}(k) \leq P_{IdB}(i,k) \leq P_{IdBLL}(k) \\ P_{OdBLL}(k) - P_{IdB}(i,k), & P_{IdB}(i,k) > P_{IdBLL}(k) \end{cases} \tag{6}$$

Block diagram of the spectral modification is shown in Fig. 3. For each frequency sample k in ith frame, the input level $P_{IdB}(i,k)$ is calculated as the sum of squared magnitude of the spectral samples in the band centered at k and with bandwidth corresponding to auditory critical bandwidth given as

$$BW(k) = 25 + 75\left(1 + 1.4(f(k))^2\right)^{0.69} \tag{7}$$

where $f(k)$ is the frequency of the kth spectral sample in kHz.

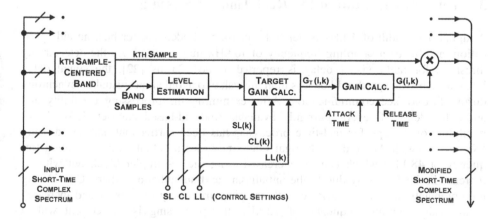

Fig. 3. Spectral modification for compensation of increased hearing thresholds and decreased dynamic range using sliding-band dynamic range compression.

For spectral modification, the target gain is converted to linear scale. The gain applied to the kth spectral sample in the ith frame is obtained using the desired attack and release times by updating the gain from the previous value towards the target value, as given in Eq. 6, and is given as

$$G(i,k) = \begin{cases} \max(G(i-1,k)/\gamma_a, G_T(i,k)), G_T(i,k) < G(i-1,k). \\ \min(G(i-1,k) \cdot \gamma_r, G_T(i,k)), G_T(i,k) \geq G(i-1,k). \end{cases} \qquad (8)$$

The number of steps during the attack and release phases are controlled using gain ratios $\gamma_a = (G_{max}/G_{min})^{1/s_a}$ and $\gamma_r = (G_{max}/G_{min})^{1/s_r}$, respectively. Here G_{max} and G_{min} are the maximum and minimum values of target gain. The number of steps during attack s_a and the number of steps during release s_r are selected to set the attack time as $T_a = s_a S/f_s$ and release time as $T_r = s_r S/f_s$, where f_s is the sampling frequency and S is the number of samples for frame shift. A fast attack avoids the output level from exceeding the uncomfortable level during transients, and a slow release avoids the pumping effect or amplification of breathing.

The processing for dynamic range compression modifies the magnitude spectrum and the signal is resynthesized using the original phase spectrum. It results in perceptible distortions due to phase discontinuities in the modified short-time complex spectrum. To mask these distortions, the analysis-synthesis method based on the least squared error estimation (LSEE) as proposed by Griffin and Lim [21] is used. It involves segmenting the input signal using L-sample frames with 75% overlap, i.e. frame shift $S = L/4$, and multiplying the segmented frames with modified-Hamming window. Complex spectrum of L-sample frame is obtained after zero-padding it to length N and calculating N-point FFT. After spectral modification as described earlier, N-point IFFT is calculated to get the modified output frame which is multiplied with the modified-Hamming window. The successive output frames are added in accordance with the overlap of the input frames to provide the resynthesized output signal.

3   App Implementation for Real-Time Processing

A signal bandwidth of 4 kHz is generally considered adequate for hearing aid application, and hence a sampling frequency of 10 kHz may be used. As the detectability threshold for audio-visual delay is reported to be 125 ms [22], the signal delay introduced by the app should be less than this value to avoid audio-visual de-synchrony during face-to-face communication. After examining the processing capability and audio I/O delay of several currently available Android-based handsets [23], implementation of the app for real-time processing has been carried out and tested using 'Nexus 5X' with Android 7.0 Nougat OS. It provides a default audio sampling frequency of 48 kHz which can be decreased using the re-sampler block but with an increase in the I/O delay due to the antialiasing and smoothening filters. Considering the computational load associated with sampling frequency of 48 kHz and increased delay with sampling frequency of 12 kHz, the processing is carried out with a down-sampling ratio of two resulting in sampling frequency of 24 kHz. The FFT-based analysis-synthesis uses 20-ms frames with 75% frame overlap and 1024-point FFT. Thus the implementation uses $f_s = 24$ kHz, $L = 480$, $S = 120$, and $N = 1024$.

The app "Hearing Assistant 1.0" has been developed with a GUI for control settings, enabling graphical setting of the controls for threshold and range compensations. The screen is used in its landscape mode to make an appropriate use of the aspect-ratio, ignoring the 'auto-rotate' setting. The screenshot of the home screen of the app is shown in Fig. 4. The play/stop button is for on/off control. The two bars at the right side graphically display the input and output levels for assessing the effect of control settings. The graphical control for gain and range compensation is accessed through the 'settings' button of 'compression' module. A provision has been made to include other processing modules, such as noise suppression [24, 25] and CVR enhancement [26] in the later versions of the app. All modules have on/off and 'settings' buttons. The on/off button can be used for toggling the function and the settings button can be used for setting the processing parameters graphically.

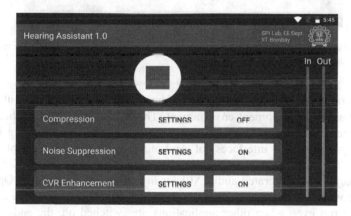

Fig. 4. Screenshot of home screen of hearing aid app "Hearing Assistant 1.0"

Figure 5 shows the screenshot of the 'settings' screen for compression. It provides touch control of points, called 'thumbs', on the level vs frequency plots for setting the output levels for soft, comfortable, and loud sounds, for up to 10 frequencies. A thumb can be moved by dragging, freely in the horizontal direction and as constrained by the upper and lower curves in the vertical direction. Movement of a thumb on one curve along x-axis moves the corresponding thumbs on the other two curves, while the movement along y-axis does not affect them. When a touch event occurs, the thumbs on the three curves closer with respect to the x-position of the touch are shifted to the new x-position, and the closest one with respect to the y-position is shifted to the new y-position. This method permits changing a thumb position without dragging it from its earlier position. With each change in the thumb position, the curves for all the intermediate frequency values are obtained by a smooth curve fitting through the thumbs, using Fritsch-Carlson method of monotonic cubic spline fitting [27] to ensure that the curve between two adjacent control points is monotonic. The gains calculated are transferred to the spectral modification block at the next analysis frame, permitting a real-time and interactive touch-control of the processing parameters. The values of input and output levels in each frame are transferred to GUI for updating the level bars.

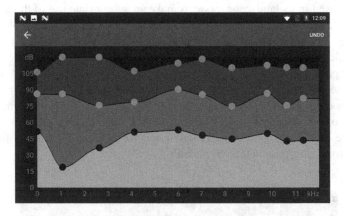

Fig. 5. Screenshot of settings screen to set parameters of frequency dependent gain and dynamic range compression

The program was written using a combination of C++ and Java, with Android Studio 2.1.0 as the development environment. The block diagram of application process of the hearing aid app is shown in Fig. 6. As the audio I/O delay using Java APIs (application programming interfaces) of Android is beyond the acceptable value of 125 ms, the I/O framework is implemented in C++ using Open SLES library [28] which is a hardware-accelerated audio API for input/output handling. An Open SLES audio engine with recording and playback capabilities is realized. Since the application is required to run even when other applications are switched on, the audio engine is maintained using a high priority background service. The service is started when the application comes to foreground and remains running until the playback is stopped or the application process is killed. The communication between GUI and the audio engine is carried out through background service using JNI (Java native interface) as shown in Fig. 6.

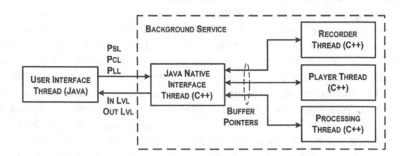

Fig. 6. Block diagram of application process in the hearing aid app.

A recorder and a player object are realized with required sampling rate and frame length specifications, each with internal queues to manage the buffer blocks currently being recorded/played. Separate threads are spawned for recorder and player to ensure uninterrupted I/O. The recorder and player threads are scheduled at least once within the frame shift duration of 5 ms. Blocking calls to the operating system (OS) such as memory allocation, mutex locks, etc. can cause underruns in real-time applications. To minimize the blocking calls, the memory required to handle I/O and processing are allocated while creating the audio engine. The processing and the signal acquisition and playback are carried out using pointers and no memory allocation is required while the app is running. The new samples from ADC are filled in the S-word free buffer of recorder thread. The recorder thread uses $2S$-word buffer queue and it passes the pointer of S-word filled buffer to JNI interface when scheduled. The processing thread when scheduled gets pointer of S-word buffer containing the new samples from the JNI interface and returns the pointer of S-word buffer containing processed samples. The player buffer when scheduled gets pointer of S-word buffer containing processed samples from JNI interface and these samples are output through DAC.

In the processing thread, the input samples, real and imaginary parts of spectral samples, and the output samples are stored as 32-bit floating point arrays. The input samples are acquired in S-word blocks from JNI interface. For each S-word just-filled block, a $3S$-word buffer stores the previous input samples. Input window with L samples ($4S$ words) is formed using the samples of the just-filled block and the samples from previous three blocks stored in $3S$-word buffer. These L samples are multiplied by L-point modified-Hamming window and N-point FFT is calculated. Open source Kiss FFT library [29] has been used for floating point FFT computation. Input power in an auditory critical band centered at each of the first $N/2$ complex spectral components is calculated as given in Eq. 7. Modified spectrum is obtained by multiplying first $N/2$ complex spectral samples with the corresponding gains. The other samples of modified spectrum are obtained by taking complex conjugate of first $N/2$ complex spectral samples. N-point IFFT of the modified complex spectrum is calculated and the resulting sequence is multiplied with the modified-Hamming window to get the modified output frame. The output signal is re-synthesized using overlap-add. In the current implementation, the processed signal is output on both the audio channels. Separate graphical settings of the processing parameters for the two channels may be incorporated for use of the app as a binaural hearing aid. The processing has algorithmic delay of one frame, i.e. 20 ms. For real-time output, the processing of each frame has to get completed before the arrival of the next frame, i.e. the frame shift interval of 5 ms, resulting in the signal processing delay, comprising the algorithmic and computational delays, of 20–25 ms.

4 Test Results

The app was tested on the handset model 'Nexus 5X' with Android 7.0 OS. Qualitative evaluation was carried out using the headset of the handset for speech input through its microphone and audio output through its earphone. The objective evaluation of the processing was carried out using an audio input-output interface with a 4-pin TRRS (tip-ring-ring-sleeve) connector to the headset port of the handset, as shown in Fig. 7. It has a resistive attenuator for attenuating the input audio signal to a level compatible with the microphone signal level and output resistance of 1.8 kΩ for it to be recognized as external microphone. The two output channels have 100 Ω load resistances. A PC sound card was used for applying the audio input and acquiring the processed output. Additional testing was carried out by providing signals from a function generator and observing the waveform using a digital storage oscilloscope.

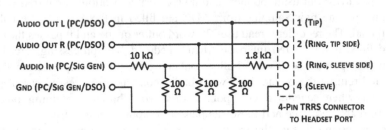

Fig. 7. Audio interface to the 4-pin TRRS headset port of the mobile handset

The settable frequency response without compression was tested using a sinusoidal input of swept-frequency and constant amplitude. An example of the processing along with the frequency-dependent gains set by the GUI is shown in Fig. 8, with the soft level kept high in bands at 1 kHz and 7.5 kHz, and the comfortable level and the loud level kept same and at the maximum value. The input frequency is linearly swept from 100 Hz to 10 kHz over 10 s. The processed output shows changes in amplitude of the sine wave in accordance with the soft-level curve. The processing for dynamic range compression was tested by setting different values for comfortable and loud levels and amplitude-modulated tones of different frequency and amplitude envelopes as applied inputs. An example of the compression is shown in Fig. 9. The input is an amplitude-modulated tone of 1 kHz. The processing gives higher gains at lower values of the input level. Spikes in the amplitude envelope of the output signal in response to step changes in the amplitude envelope of the input signal, as seen in the figure, are typical of the dynamic range compression with a finite frame shift and can be eliminated by using one-sample frame shift but with a significantly increased computation load.

The app was tested for speech modulated with different types of amplitude envelopes. An example of the processing is shown in Fig. 10, with an English sentence repeatedly concatenated with different scaling factors to observe the effect of variation in the input level on the output waveform. Informal listening test was carried out with different speech materials, music, and environmental sounds with large variation in the sound level as inputs. The outputs exhibited the desired amplification and compression without introducing perceptible distortions. There were no perceptible differences between the processed outputs from the app and the offline implementation. Examples of the processing by the app are available at [30].

The audio latency, i.e. the signal delay comprising the processing delay and I/O delay, was measured by applying a 1 kHz tone burst of 200 ms from a function

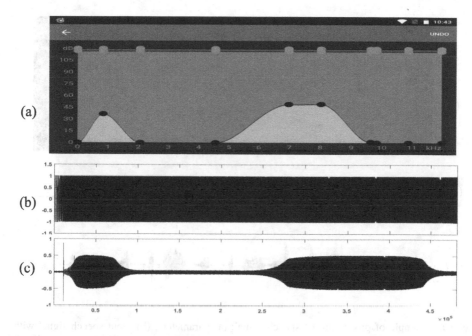

Fig. 8. Example of processing for frequency-dependent linear gain: (a) parameters, (b) input of tone of constant amplitude with frequency linearly swept from 100 Hz to 10 kHz over 10 s, and (c) processed output

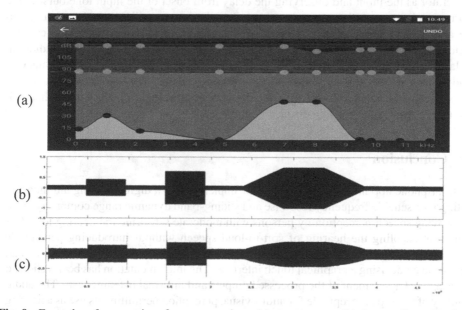

Fig. 9. Example of processing for compression: (a) parameters, (b) input of amplitude-modulated sinusoidal tone of 1 kHz, and (c) processed output

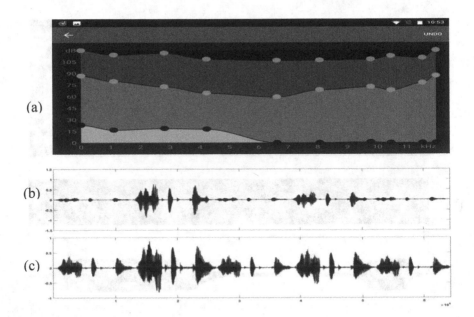

Fig. 10. Example of processing of speech signal: (a) parameters, (b) input speech signal with amplitude modulation (English sentence "*you will mark ut please*" repeatedly concatenated with different scaling factors), and (c) processed output

generator as the input and observing the delay from onset of the input tone burst to the corresponding onset in the output, using a digital storage oscilloscope. It was found to be approximately 55 ms. The processing delay (sum of the algorithmic and computational delays) is 1.25 times the frame length, i.e. approximately 25 ms. The additional delay is due to audio I/O delay of the handset hardware, buffering operations in the OS, and delays in the anti-aliasing and smoothening filters. The audio latency of our implementation is much lower than the detectability threshold of 125 ms and hence it should be acceptable for face-to-face communication.

5 Conclusion

A smartphone app has been developed for implementing a digital hearing aid incorporating user settable frequency response and sliding-band dynamic range compression for improving speech perception by persons with moderate sensorineural hearing loss. It is aimed at enabling the hearing of soft-to-loud speech without introducing perceptible distortions. The processing parameters can be set by the user in an interactive and real-time mode using a graphical touch interface. The implementation has been validated by graphical assessment of the processed outputs and informal listening tests. The audio latency of the app is acceptable for audio-visual perception permitting its use as a hearing aid during face-to-face conversation. Use of the app by a large number of hearing impaired listeners is needed for its real life evaluation and further enhancement.

Interactive settings and processing for binaural hearing aid may be incorporated in the app. It may be extended as a hearing aid for persons with severe loss by using headsets with appropriate output levels. With further increase in the processing capacity and decrease in audio I/O delay of newer handset models, there is scope for incorporating processing for speech enhancement in the app for extending its usability in noisy environments.

Acknowledgment. The research is supported by "National Program on Perception Engineering Phase-II" sponsored by the Department of Electronics & Information Technology, Government of India.

References

1. Levitt, H., Pickett, J.M., Houde, R.A. (eds.): Sensory Aids for the Hearing Impaired. IEEE Press, New York (1980)
2. Moore, B.C.J.: An Introduction to the Psychology of Hearing, pp. 66–107. Academic, London (1997)
3. Gelfand, S.A.: Hearing: An Introduction to Psychological and Physiological Acoustics, 3rd edn, pp. 314–318. Marcel Dekker, New York (1998)
4. Dillon, H.: Hearing Aids. Thieme Medical, New York (2001)
5. Sandlin, R.E.: Textbook of Hearing Aid Amplification. Singular, San Diego, pp. 210–220 (2000)
6. Byrne, D., Tonnison, W.: Selecting the gain of hearing aids for persons with sensorineural hearing impairments. Scand. Audiol. **5**, 51–59 (1976)
7. McCandless, G.A., Lyregaard, P.E.: Prescription of gain/output (POGO) for hearing aids. Hear Instrum. **34**(1), 16–21 (1983)
8. Braida, L.D., Durlach, N.I., Lippmann, R.P., Hicks, B.L., Rabinowitz, W.M., Reed, C.M.: Hearing aids – a review of past research on linear amplification, amplitude compression, and frequency lowering. In: Hardy, J.C., (ed.) ASHA Monographs, Rockville, MA (1979). www. asha.org/uploadedFiles/publications/archive/Monographs19.pdf
9. Lippmann, R.P., Braida, L.D., Durlach, N.I.: Study of multichannel amplitude compression and linear amplification for persons with sensorineural hearing loss. J. Acoust. Soc. Am. **69**, 524–534 (1981)
10. Asano, F., Suzuki, Y., Sone, T., Kakehata, S., Satake, M., Ohyama, K., Kobayashi, T., Takasaka, T.: A digital hearing aid that compensates for sensorineural impaired listeners. In: Proceedings of IEEE International Conference on Acoustics, Speech, and Signal Processing 1991 (ICASSP 1991), Toronto, Canada, pp. 3625–3628 (1991)
11. Stone, M.A., Moore, B.C.J., Alcántara, J.I., Glasberg, B.R.: Comparison of different forms of compression using wearable digital hearing aids. J. Acoust. Soc. Am. **106**, 3603–3619 (1999)
12. Tiwari, N., Pandey, P.C.: A sliding-band dynamic range compression for use in hearing aids. In: Proceedings of 20th National Conference on Communication (NCC 2014), Kanpur, India (2014). doi:10.1109/NCC.2014.6811300
13. Pandey, P.C., Tiwari, N.: Dynamic range compression with low distortion for use in hearing aids and audio systems, International PCT Application number PCT/IN2015/000049 (2015)

14. Tiwari, N., Pandey, P.C., Sharma A.: A smartphone app-based digital hearing aid with sliding-band dynamic range compression. In: Proceedings of 22nd National Conference on Communication (NCC 2016), Guwahati, India (2016) doi:10.1109/NCC.2016.7561146
15. Mimi Hearing Technologies: Mimi hearing test v3.0. itunes.apple.com/in/app/mimi-hearing-test/id932496645
16. IT4You: Petralex hearing aid v1.4.3. play.google.com/store/apps/details?id=com.it4you.petralex, http://itunes.apple.com/us/app/petralex-hearing-id
17. Clark, N.: Bio aid v1.0.2. itunes.apple.com/gb/app/bioaid
18. Lamberg Solutions: Hearing aid with replay v2.0.0. play.google.com/store/apps/details?id=com.ls.soundamplifier
19. Newbrick, S.A.: uSound (hearing assistant) v.1.6. play.google.com/store/apps/details?id=com.newbrick.usound
20. Quadio Devices: Q+hearing aid. play.google.com/store/apps/details?id=com.quadiodevices.qplus
21. Griffin, D.W., Lim, J.S.: Signal estimation from modified short-time Fourier transform. IEEE Trans. Acoust. Speech Signal Process. **32**, 236–243 (1984)
22. International Telecommunication Union: Relative timing of sound and vision for broadcasting, ITU Rec. ITU-R BT.1359 (1998)
23. Superpowered Inc: Android audio latency and iOS audio latency test app superpowered.com/latency
24. Loizou, P.C.: Speech Enhancement: Theory and Practice. CRC, New York (2007)
25. Tiwari, N., Pandey, P.C.: Speech enhancement using noise estimation based on dynamic quantile tracking for hearing impaired listeners. In: Proceedings of 21st National Conference on Communication (NCC 2015), Mumbai, India (2015). doi:10.1109/NCC.2015.7084849
26. Jayan, A.R., Pandey, P.C.: Automated modification of consonant-vowel ratio of stops for improving speech intelligibility. Int. J. Speech Technol. **18**, 113–130 (2015)
27. Fritsch, F.N., Carlson, R.E.: Monotone piecewise cubic interpolation. SIAM J. Numer. Anal. **17**(2), 238–246 (1980)
28. Khronos group: OpenSLES. www.khronos.org/opensles/
29. Borgerging, M.: Kiss FFT. sourceforge.net/projects/kissfft
30. Sharma, S.: Dynamic range compression results from smartphone app implementation (2016). www.ee.iitb.ac.in/~spilab/material/saketh/ihci2016

Hands Up! To Assess Your Sustained Fitness

Ankita Singh, Richa Tibrewal, Chandrima Bhattacharya,
and Malay Bhattacharyya[✉]

Department of Information Technology,
Indian Institute of Engineering Science and Technology,
Shibpur, 711103, West Bengal, India
ankitasinghchikki@gmail.com, richa.besu@gmail.com, chandrima.004@gmail.com,
malaybhattacharyya@it.iiests.ac.in

Abstract. A better understanding of the mobile sensors and the need
for mobile information management have improved the personal healthcare in recent years. In this study, we present a mobile-healthcare
approach for the automatic assessment of sustainable fitness in hands
correlated to different diseases. Arm weakness is a common symptom in
many of the chronic diseases like arthritis or cardiac arrest. It is also
a prominent signature of verifying fitness in many healthy persons too.
To detect the amount of weakness in arm, we perform a simple test of
lifting hand using a smart phone. With this approach, we can basically
quantify the fitness of arm and routinely track whether the condition of
the subject sustains or not.

Keywords: Mobile healthcare · Sensors · Assisted and enhanced living

1 Introduction

The use of smartphones for mobile-healthcare has received immense interest of
researchers from diverse fields in recent years [1,2,11]. In fact, the use of smartphones has radically changed healthcare support in the recent decade [3,12].
Testing the sustained physical ability of a user exploiting a smartphone is one
such promising area of research. In this study, we propose an assistive smartphone application (app) for the persons with disabilities. Its working principle
resembles the procedure of medical assessment for arm weakness. Smartphone
apps are already in use for a couple of years to monitor the health parameters of
a person. Very recently, smart watches have also been introduced which employ
intelligent apps taking various health based parameters as inputs. They collect
the health-related data like pulse rate, total amount of calories burned, total
distance traversed by running, and other such relevant information. Many of the
apps make a clever use of these data for accurate prediction of occurrence of
diseases. We also exploit the power of smart sensors and make their intelligent
use for a better processing of health-related data. The purpose is to efficiently
verify sustained fitness of both the hands of a given subject.

© Springer International Publishing AG 2017
A. Basu et al. (Eds.): IHCI 2016, LNCS 10127, pp. 187–194, 2017.
DOI: 10.1007/978-3-319-52503-7_15

2   Motivation

Several chronic diseases have been reported earlier to have a connection with the problem in lifting hands over the head. This includes asthma [8], respiratory problems [7], arthritis [14], amnesia [4], stroke [6,9], crossed cerebellar diaschisis [10], parakinesia brachialis oscitans [13], etc. The relevant symptoms of these diseases associated to the fitness of arm, which we aim to detect in this study, are listed in Table 1. Our motivation is to recognize arm weakness in real-time and take automatic actions accordingly. This could be sending attention messages to the emergency contacts or services as and when required. Additionally, we wish to quantify the strength of hand to decide the sustainability of fitness of a (disable, as well as normal) person.

Table 1. The diseases having a direct or indirect effect on the strength of hand, thereby creating problems while lifting hands over the head.

Disease	Effect on hand
Asthma	Weakness of both the arms
Respiratory problems	Weakness of both the arms
Arthritis	Weakness of both the arms
Amnesia	Arm specific weakness
Stroke	Incapability of lifting
Crossed cerebellar diaschisis	Incapability of lifting
Parakinesia brachialis oscitans	Involuntary lifting

3   Related Works

The rehabilitation of patients is an important issue in medical research [5]. With the development of information technology, it has become imperative to keep track of the health related data of recovering patients for their better monitoring. In fact, monitoring the performance of different limbs has been shown to be effective for a better treatment of rehabilitating patients [4,5]. Several studies have been carried out in this direction that show the therapeutic importance of regular health-monitoring [7].

Emergence of smartphones and the use of cheap (but effective) sensors within it have boosted the development of apps that target both patients and normal people for healthcare support [3]. These mobile based health-monitoring methods have been proved to significantly improve self-monitoring physical activities and the study of symptoms related to several diseases. Some topical studies also highlight the advantages of using mobile apps for a better treatment of patients [12]. There are developments in recent years that aims specific diseases with smartphone based apps [11]. However, no previous study has been carried out aiming the self-rehabilitation of patients using smartphone apps. In this paper, we present an app for the assessment of sustained fitness of arms of patients of various chronic diseases.

4   Recognizing the Weakness of Hand

Problem in lifting both the hands ensures whether there is any weakness, may be partial or complete, in the arm. The app basically asks the user to hold the phone facing the screen on the top and lift up his hand (as shown in Fig. 1(a)). To ensure a correct position, an alarm tone is rung. Promptly after this, the user is required to lift up his hands (as shown in Fig. 1(b)) until the smartphone comes to a vertical position and an alarm tone is heard to confirm the position. The phone is hold in an unusual (in fact strange) way just to make a guarantee that the phone has not simply been rotated around the Z-axis to attain a vertical position. The touchscreen is used to verify the position of both palms on the phone.

We acquire the accelerations along X-, Y- and Z-axis measured by the accelerometer sensor present in the smartphone as main inputs. The accelerometer is used to measure the acceleration forces along the different axes of the smartphone with considerable accuracy. These accelerations are the net acceleration values along each direction, i.e., accelerations due to gravity as well as the other dynamic forces on the smartphone like external movements, vibration, etc.

As can be seen from Fig. 1, the acceleration values are different at the end points of the smartphone while lifting a hand. The two end points of this lifting operation are expected to sense the acceleration values as shown in Fig. 1. When the subject raises his hand, the acceleration along the Y-axis gradually increases from 0 to g (corresponding to the horizontal and vertical positions, respectively). On the other hand, the acceleration along the Z-axis gradually decreases from g to 0 (corresponding to the horizontal and vertical positions, respectively). This combined pattern is used to recognize whether the subject has been able to lift his hand completely or not.

Based on the acceleration values obtained as input, the arm weakness can be quantified in terms of the angle (to the horizontal axis) the hand able to raise the phone. If it is a complete raise then the angle is $\frac{\pi}{2}$ (see Fig. 2). As we know that the net acceleration along the Y-axis is due to the effect of g, we can write the following relation for a raise of angle θ.

$$g\sin\theta = A \Rightarrow \theta = \sin^{-1}(A/g).$$

Therefore, we can quantify the level of fitness ($\mathcal{F} \in [0,1]$) of the hand (of a given subject) as follows.

$$\mathcal{F} = \frac{2\sin^{-1}(A/g)}{\pi}.$$

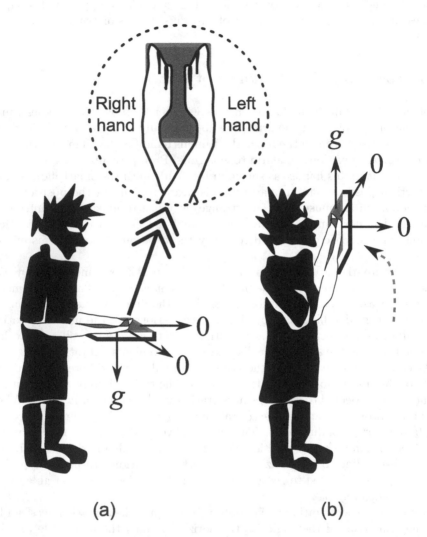

Fig. 1. The acceleration values are measured by a 3-axis accelerometer present in a smartphone while the hand lifts up the phone (a) from horizontal position facing the screen up (the acceleration values along the X-, Y- and Z-axis are 0, 0 and g, respectively), (b) to the vertical position keeping the phone straight (the acceleration values along the X-, Y- and Z-axis are 0, g and 0, respectively). The phone is required to be hold with the hands crossed and palms facing down so that the last two fingers touch upon the screen (as highlighted). To hold the phone in this way, thumbs are placed at the back of phone. The phone is required to be taken in a static position at both the end points.

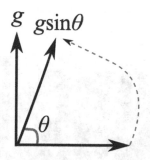

Fig. 2. The effect on net acceleration along the Y-axis due to lifting the hand.

This normalized measure can be used to recognize the arm weakness and the corresponding fitness details of a user can be stored for a routine checkup. For monitoring the condition of arm (and verifying whether it is improving or not for specific cases) over a period, this fitness measure can be stored in the smartphone itself.

A simple test can be performed to make a decision whether the strength of arm is deteriorating or not. The outcome of this test is automatically decided by the app. After the decision is taken, an alert SMS can be sent to the emergency contact, if required. In this way, mobile-based persuasive technologies can be used for a sustainable health monitoring.

5   Empirical Analysis

The environment used for empirical studies and details about the experiments carried out are given hereunder.

5.1   Configuration Details

The experiments are run on a smartphone having 1 GB RAM, Qualcomm Snapdragon 400 processor with 1.2 GHz quad-core CPU, and Android version 5.1. The smartphone uses a 3-axis accelerometer sensor. This configuration is minimal and available in any standard smartphone.

5.2   Results

To test the performance of the app, it was initially connected to a stroke recognition model mSTROKE [11], for collecting real-life data. We received data from almost 10 subjects, with an average age of 30, regarding their responses to the app. To be honest, neither of them were actual stroke patients nor having any chronic disease, however, they responded like patients with arm weakness after proper training. Note that, the app is inherently measuring the fitness of both

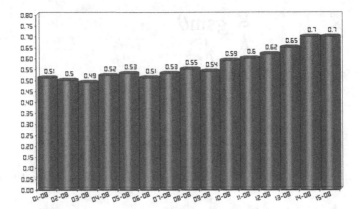

Fig. 3. A snapshot of arm fitness values that are calculated for a time period of last 15 days are shown on demand.

the arms. The test results highlight that their arm fitness was recognized appropriately in approximately 90% of the cases. The analysis was done over a period of two weeks.

A collection tested fitness values, which are shown on the mobile screen, are highlighted in Fig. 3. It is understandable that such backup details will help the user to self-assess the condition of arm and its sustainability.

6   Challenges

The sensor values of any smartphone are highly sensitive even to the slightest movement (or vibration) of the phone. Therefore, a tolerance level is required to be set while learning from the acceleration patterns. However, setting an appropriate tolerance level for this purpose is a challenging task. Although the described app has been made for a general purpose for common people, we need to leave out the users who are completely physically disabled, e.g., the persons who cannot even lift their hands due to some diseases. The way we perform the test is not applicable to them. This app might not also work for the deaf people who cannot confirm the position of their phones by listening to the alarm tone.

7   Discussion

In this paper, we describe a health-monitoring app that takes input from the simple hand movements of an user and can determine whether the user has any arm weakness or not. This smartphone based attention mechanism might be helpful to routinely check the sustainability of hand fitness of a physically challenged person. As the operating temperature of an accelerometer in a smartphone typically ranges between $[-20, 70]\,°C$ (or more) and is quite accurate, it can be universally used for the said purpose. As a limitation, this health-monitoring app is not able to distinguish the weakness in individual arms, which might be useful for rehabilitation of some specific diseases [4].

Acknowledgments. The work of Malay Bhattacharyya is supported by the Visvesvaraya Young Faculty Research Fellowship 2015-16 of DeitY, Government of India. The first two authors contributed equally to this work.

References

1. Abidi, S. (ed.): Mobile Health: A Technology Road Map. Springer, New York (2015). doi:10.1007/978-3-319-12817-7
2. Bhattacharyya, M.: Disease dietomics. ACM XRDS **21**(4), 38–44 (2015) http://dx.doi.org/10.1145/2788508
3. Boulos, M.N.K., Wheeler, S., Tavares, C., Jones, R.: How smartphones are changing the face of mobile and participatory healthcare: an overview, with example from eCAALYX. Biomed. Eng. OnLine **10**(1), 1 (2011) http://dx.doi.org/10.1186/1475-925X-10-24
4. Carpenter, K., Berti, A., Oxbury, S., Molyneux, A.J., Bisiach, E., Oxbury, J.M.: Awareness of and memory for arm weakness during intracarotid sodium amytal testing. Brain **118**(1), 243–251 (1995). http://dx.doi.org/10.1093/brain/118.1.243
5. Fardy, P.S., Webb, D., Hellerstein, H.K.: Benefits of arm exercise in cardiac rehabilitation. Phys. Sportsmed. **5**(10), 30–41 (1977) http://dx.doi.org/10.1080/00913847.1977.11710641
6. Harbison, J., Hossain, O., Jenkinson, D., Davis, J., Louw, S.J., Ford, G.A.: Diagnostic accuracy of stroke referrals from primary care, emergency room physicians, and ambulance staff using the face arm speech test. Stroke **34**, 71–76 (2003). http://dx.doi.org/10.1161/01.STR.0000044170.46643.5E
7. Jonghe, B.D., Bastuji-Garin, S., Durand, M.-C., Malissin, I., Rodrigues, P., Cerf, C., Outin, H., Sharshar, T.: Groupe de Reflexion: respiratory weakness is associated with limb weakness and delayed weaning in critical illness. Crit. Care Med. **35**(9), 2007–2015 (2007). http://dx.doi.org/10.1097/01.ccm.0000281450.01881.d8
8. McFadden Jr., E.R., Ingram Jr., R.H.: Exercise-induced asthma - observations on the initiating stimulus. New England J. Med. **301**, 763–769 (1979). http://dx.doi.org/10.1056/NEJM197910043011406
9. Nor, A.M., McAllister, C., Louw, S.J., Dyker, A.G., Davis, M., Jenkinson, D., Ford, G.A.: Agreement between ambulance paramedic-and physician-recorded neurological signs with Face Arm Speech Test (FAST) in acute stroke patients. Stroke **35**(6), 1355–1359 (2004). http://dx.doi.org/10.1161/01.STR.0000128529.63156.c5
10. Pantano, P., Baron, J.C., Samson, Y., Bousser, M.G., Derouesne, C., Comar, D.: Crossed cerebellar diaschisis: further studies. Brain **109**, 677–694 (1986). http://dx.doi.org/10.1093/brain/109.4.677
11. Tibrewal, R., Singh, A., Bhattacharyya, M.: mSTROKE: a crowd-powered mobility towards stroke recognition. In: Proceedings of the 18th International Conference on Human-Computer Interaction with Mobile Devices and Services, Florence, Italy, 6–9 September, pp. 645–650. ACM Press (2016). http://dx.doi.org/10.1145/2957265.2961831

12. Turner-McGrievy, G.M., Beets, M.W., Moore, J.B., Kaczynski, A.T., Barr-Anderson, D.J., Tate, D.F.: Comparison of traditional versus mobile app self-monitoring of physical activity and dietary intake among overweight adults participating in an mHealth weight loss program. J. Am. Med. Inform. Assoc. 20(3), 513–518 (2013). http://dx.doi.org/10.1136/amiajnl-2012-001510

13. Walusinski, O., Neau, J.-P., Bogousslavsky, J.: Hand up! Yawn and raise your arm. Int. J. Stroke 5(1), 21–27 (2010). http://dx.doi.org/10.1111/j.1747-4949.2009.00394.x

14. Zeidman, S.M., Ducker, T.B.: Rheumatoid arthritis: neuroanatomy, compression, and grading of deficits. Spine 19(20), 2259–2266 (1994). http://dx.doi.org/10.1097/00007632-199410150-00003

A Voting-Based Sensor Fusion Approach for Human Presence Detection

Sonia$^{(\boxtimes)}$, Manish Singh, Rashmi Dutta Baruah,
and Shivashankar B. Nair

Indian Institute of Technology Guwahati, Guwahati 781039, Assam, India
{s.sonia,manish.singh,r.duttabaruah,sbnair}@iitg.ac.in

Abstract. With advances in technologies, environments and human habitats are all set to become smarter. Such environments, which include smart homes, hospitals, campuses, etc., are oriented towards human comfort and safety. Detecting the presence of a human being in spaces within such environment forms a major challenge. Though there are sensors that can perform this task, they are not without limitations. Using information derived from either single or multiple sensors separately is not sufficient to distinguish human beings from other objects within the environment. Reliable detection of human presence can be achieved only by fusing information obtained from multiple sensors. In this paper, we describe an approach to human presence detection using a combination of PIR and ultrasonic sensors. Analysis is performed based on the received raw signals from the PIR as well as an analog ultrasonic sensor. A voting based approach has been used to classify signals obtained from human beings and non-human objects, thereby facilitating human presence detection. Results obtained from indoor experiments performed using this approach substantiate the viability of its use in real environments.

Keywords: Human sensing · Sensor fusion · Voting-based approach

1 Introduction

Research in the area of smart environments is constantly striving towards improving the ecosystem, which we inhabit. For e.g. one of the objectives of a smart home is to provide comfort to elderly people. Such homes allow them to live independently and perform daily activities effortlessly. Apart from providing comfort, the other requirements from a smart home are sustainability and safety. The former can be achieved through smart energy management systems while the latter can be met by smart security systems. Since human presence is an integral factor within these environments, the embedded systems responsible for making them smart rely heavily on human presence detection. Detecting human presence is still an open problem and presents a big challenge. A human being can be considered to be very dynamic in the sense that s/he can assume numerous body

© Springer International Publishing AG 2017
A. Basu et al. (Eds.): IHCI 2016, LNCS 10127, pp. 195–206, 2017.
DOI: 10.1007/978-3-319-52503-7_16

postures, walk at varying speeds, wear a range of widely varying clothing material and even think of ways to escape detection [26]. The complexity of detecting human presence increases when we take into consideration the aspect of predicting human behavior. Applications like intruder detection, elderly fall detection, health care applications, human tracking, etc. solely depend on the detection of human beings within the given environment. Based on the application at hand, researchers have used a variety of sensors for human presence detection. Vision based sensors are one of the most commonly used sensors for human detection [4,12]. In each of these cases the associated computations that include image segmentation, morphological operations and image filtering, are computationally heavy and thus time consuming. Since illumination is mandatory, these techniques cannot work in the dark. These sensors however are unsuitable for applications wherein privacy is a matter of concern. Smart homes, where a fall inside a washroom needs to be detected, fall under this category. Vision sensors also fail to serve their purpose in rainy or dusty environments. Such sensors can also be fooled into detecting idols, mannequins or a robot, as human beings. E. Machida et al. have used the Kinect sensor [15] to detect the presence of a human being. Though the technique described therein is effective, the cost of the gadget and the related computations involved does not make it a practically viable solution. Laser sensors have been used by Katsuyuki Nakamura et al. [20] for human detection. The outline of the human being is used to differentiate humans from non-humans. The use of lasers may however pose health issues. PIR sensors, that work on temperature difference, have been used for sensing human presence. Unlike, other sensors (such as Ultrasonic, Laser, etc.), a PIR sensor does not bombard any kind of radiations on the obstacle in front. It works by detecting the heat energy (infrared radiations) emitted or reflected by various objects in the surroundings. Soyer et al. [27] present the benefits of a PIR sensor for indoor event detection when used in conjunction with security in an office or home. M. Moghavvemi et al. [17] have used a PIR sensor to detect intruder within a room. Similar work has been cited in [8] by using different algorithms with the same sensor. Moghavvemi et al. [17] have developed an indoor awareness system for smart homes which relies on a PIR sensor. However, the system fails to differentiate between animals (e.g. pets in a home) and human beings. PIR sensors also fail in presence of hot objects. The authors claim that detection can be performed on condition that the human being is in motion. B. Song et al. [25] describe a surveillance tracking system using multiple PIR sensors. The algorithm used for tracking, is region based and the system works even in dark regions, where a conventional vision based sensor fails to locate a human being. But system fails to track when human becomes stationary somewhere for long interval of time. Dong et al. [7] have shown that energy consumption can be managed by switching the lights on or off based on human occupancy which is detected using PIR sensors.

Ultrasonic sensors have been used to classify human beings from non-human ones [26]. However, just like the others, these sensors too fail to provide the desired high true positive rate (TPR) and low false positive rate (FPR). Ultrasonic sensors are mainly used for obstacle avoidance in the robotic domain.

They have also found use in recognizing human activity. The main concept behind its effective use lies in the heterogeneity of the reflected wave. They have also been used to detect the defects in machine parts [24]. These sensors have also been helpful in human face detection [29] in a limited way. Researchers have been known to use a range of other sensors for human detection which include among others, acoustic [5], etc. Unfortunately, all sensors have their own limitations and fail to detect human beings under some circumstances. One may conclude that given all kinds of scenarios, a single sensor proves to be inadequate to sense the presence of a human being. Sensor fusion [28] is a possible alternative, which researchers are now exploring [1,5]. This paradigm uses a combination of sensors, such as a PIR and a camera [6], PIR and a vibration sensor [31], etc. and interprets the data received from each of them to eventually conclude on a detection. Several applications exist where either only PIR or a combination of PIR and other sensors have been used for human detection. These include surveillance [2], positioning [9], etc. Ahmet Yazar et al. [30] have made use of a combination of PIR and vibration sensors, while Nadee et al. [19] have used infrared and ultrasonic sensors to detect the fall of elderly persons. Home appliances have been automated based on task activity detection of human beings using multiple image sensors in [21]. The use of vibration sensors is however infrastructure dependent [31] and thus can be used only in some specific scenarios. On the other hand, ultrasonic sensors are not dependent on infrastructure. The better option possibly is to combine the use of both PIR and ultrasonic sensors. As can be observed, both PIR and ultrasonic sensors have their respective and distinct advantages and limitations. This means that their combined use can result in the realization of a better system. Applications targeting smart homes [13,14], surveillance systems [3], etc., all use a combination of ultrasonic and PIR sensors. Apart from choosing a suitable combination of sensors, the selection of an appropriate classification technique to suit the application at hand is crucial. Maslov et al. [16] present the management of sensory data for real time analysis of the environment, wherein they use Dempster-Shafer theory [18], fuzzy rules [1], Bayesian Belief Network, [23], etc. techniques to fuse the data from the sensors. In order to classify the data among different classes they use several algorithms which include Fuzzy Rule based classifiers [26], Support Vector Machines [10], etc. In this paper, we describe a computationally simple and easy voting based classifier that can aid in detecting an object to be human or otherwise. Three sensors viz. a PIR, an analog ultrasonic sensor and a PING sensor have been used to detect the presence of human beings in an indoor setup. The results obtained offer great promise in the practical viability of this approach.

The subsequent section discusses methodology used while the ones that follow describe the experiments conducted along with results and performance. Conclusions arrived at and future scope for research, comprise the last sections.

2   Methodology

2.1   Voting-Based Sensor Fusion and Detection

This approach is used to fuse the sensory data from different sensors so as to aid the detection of a human being. To apply this approach, it is assumed that all the sensors point to the same target at the same instant of time and that data obtained from all the sensors is buffered at the same speed. Step 1 to step 3 of methodology explains the model development process whereas, step 4 to step 6 depicts the detection process.

Step1: Order the sensors $S_1, S_2, ..., S_N$ on the basis of their reliability in detecting human presence. Here, S_1 has the highest reliability (in terms of human detection) while S_N has the least. Data from the sensor with highest reliability is considered first for processing. Define an upper threshold τ_k for each sensor S_k (k= 1 to N). This threshold is determined experimentally and indicates the probability of an object being classified as a human being.

Step2: Data captured from a sensor S_k for a period T can be conceived to be a signal. M features are extracted from each of a total of G signals. x_{ij}^k is defined as the jth feature extracted from i^{th} signal which is received from k^{th} sensor. Thus, for all X^k($k = 1, 2, 3, ..., N$) for sensor k is represented as a matrix X^k as given below. The matrix X^k is used as training data for developing the model.

$$X^k = \begin{bmatrix} x_{11}^k & x_{12}^k & \cdots & x_{1M}^k \\ x_{21}^k & x_{22}^k & \cdots & x_{2M}^k \\ \cdot & \cdot & \cdots & \cdot \\ \cdot & \cdot & \cdots & \cdot \\ \cdot & \cdot & \cdots & \cdot \\ x_{G1}^k & x_{G2}^k & \cdots & x_{GM}^k \end{bmatrix} \tag{1}$$

Step3: Let,Y_j^k be a vector which is represented as:

$$Y_j^k = \left\{ x_{11}^k \; x_{21}^k ... x_{G1}^k \right\} \text{ where, } j = 1 \text{ to } M$$

Apply mean-shift clustering technique [11] for each Y_j^k. The mean-shift is a non-parametric method and can be used to find the maxima of a density function. In this technique a single parameter called the bandwidth is used which is internally determined by the median of all the pairs of elements of Y_j^k ($j = 1$ to M). The proximities are measured using the Gaussian method. We have used the flat kernel [22] to calculate the kernel density function. This function is used to determine all the local maxima that are cluster centers.

Applying mean shift results in a set of clusters along with their corresponding centers and radii for each Y_j^k. Thus for each sensor S_k, there will be M sets of such clusters. Let Q_j^k represents the set of centers and radius of the clusters of Y_j^k.

$Q_j^k = \{(c_{1j}^k, r_{1j}^k)(c_{2j}^k, r_{2j}^k) \dots (c_{Lj}^k, r_{Lj}^k)\}$ where, c_{pj}^k is the pth center of the pth cluster of Y_j^k, r_{pj}^k is the pth radius of the respective cluster of Y_j^k, L is the number of clusters obtained.

Step4: Let z^k represent all the M features extracted from a test signal captured via the sensor S_k when, an obstacle is in front of the sensors.

$$Z^k = \{z_1^k\, z_2^k \dots z_M^k\}$$

It may be noted that here we are taking just one test signal unlike what was done in step 2 where G number of signals were considered. In the detection process, sensor signals are processed in the order as defined in step1. So, the first sensor S_1 we obtain the following cluster centers and radii corresponding to vector Y_j^1 (according to step 3).

Thus, $Q_j^1 = \{(c_1^1, r_1^k)(c_2^1, r_2^1) \dots (c_L^1, r_L^1)\}$

Now, votes are calculated based on the membership of each z_j^1 to the respective clusters. If z_j^1 belongs to any one of the clusters associated Q_j^1, then the vote for the feature j, of sensor S_1, $V_j^1 = 1$, else $V_j^1 = -1$ (In general, the vote for feature j extracted from signal of sensor S_k is represented as Vj^k). Similarly, votes for all the M features are calculated.

Step5: Count the number of positive votes and calculate the probability of the object to be classified as human being using Eq. 2.

$$Probability(P^k) = \frac{\sum_{j=1}^{M}(V_{j=1})}{M} \tag{2}$$

Step6: If the calculated probability (P^k) adheres to the threshold (τ_k) condition for the sensor S_k then the confronted object is detected as a human being. If, the calculated probability (P^k) does not satify this threshold condition, then the step 4 through step 6 are repeated for the signal received from the next sensor.

3 Experiments and Results

The voting based sensor fusion approach is applied to fuse the data from ultrasonic and PIR sensors to enable human detection. The following hardware was used in the experiments conducted:

1. Ultrasonic Ping Sensor 2. PIR Sensor 3. Maxbotix Analog Ultrasonic Sensor (AUS) and 4. Arduino Mega 2086 board. A PIR sensor works on the temperature difference and was used to sense the human motion in front of the sensor while the ultrasonic PING sensor was used to measure the distance of the obstacle confronted. The Maxbotix Analog ultrasonic sensor was used to capture the raw ultrasonic signal reflected from the obstacle in front. The Arduino Mega was used as the interfacing device to read the sensory data through a serial port. The experiments performed were based on the analysis of raw signals received from

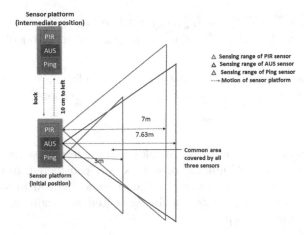

Fig. 1. Sensing range of the different sensors used

the PIR sensor as well as the AUS. As mentioned earlier the PIR sensor fails to detect the human being when s/he is stationary. This issue was resolved by analyzing the signal received from the AUS. However, for some objects the difference between two received ultrasonic signals for two different objects was found to be marginal which made it difficult to differentiate the two objects. This difference is also affected by noise. It was observed that merely on the basis of the AUS, one cannot differentiate the human being from other objects. To overcome the limitations of both the PIR sensor and the AUS, the data received from the sensors was combined before the final decision of classification was made. Instead of using these sensors separately, sensors can be used simultaneously to improve the accuracy (defined later by Eq. 11) and to reduce false alarms. A false alarm is said to be generated when a non-human object is classified as a human being. In the experiment, the three sensors were mounted on a moving platform in such a way that the sensing range of the PIR sensor overlapped that of the AUS and the PING sensor as shown in Fig. 1. The platform was made to move to and fro along a 10 cm straight line. The reason behind doing so was to sense the same target with all sensors at the same instance of time. Since the associated data sheets mention that the AUS could provide erroneous results at distances less than 50 cm from the sensor, data received from the sensors was processed only when the target was detected 60–70 cm away. The distance of the obstacle was continuously checked via data received from the PING and the analog ultrasonic sensors. It was observed that when the distance of the target was greater than 80 cm, the signal strength from the AUS was too small to be read by the Arduino board. In the experiments we have thus tried to detect the human being within the range of 60–70 cm. This is a fair distance considering the fact that a robot needs to perform the detection. As soon as an obstacle was confronted within this specified range, the data received from the sensors was processed.

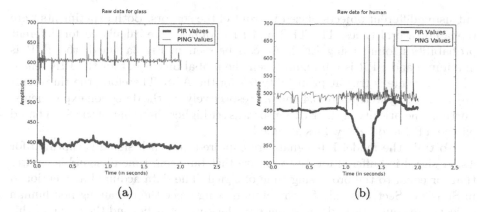

Fig. 2. Raw signals for different objects

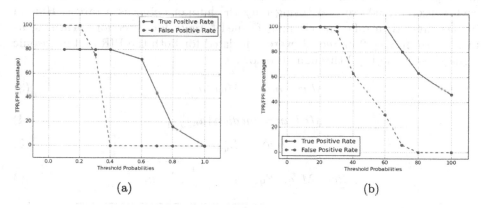

Fig. 3. (a) Performance of PIR sensor with varying probabilities (b) Performance of AUS sensor with varying probabilities

Data Acquisition: Data acquired from various sensors was buffered at 115200 bits per second (bps). As mentioned earlier, the data was buffered only if the object was in the range of 60–70 cm.

Figure 2 shows the nature of the raw signals (amplitude vs. time) received from the AUS and PIR sensors when used to detect various objects including a human being. While there is a marked visual difference between the PIR signals obtained from human and non-human objects the same is not true for those from the AUS. As described in Step 1, the order and thresholds for the PIR sensor data (τ_{PIR}) and AUS (τ_{AUS}) need to be defined. In order to set the thresholds for both the PIR (τ_{PIR}) and AUS (τ_{AUS}) sensors, two experiments viz. Experiment 1 and Experiment 2 were performed. Experiment 1 was performed with a PIR sensor for human detection. The same task was also performed with a AUS in Experiment 2. Assuming different probabilities the values of TPR and FPR were found by conducting experiment 1 and 2. Both experiments were carried out using different objects placed in front of the sensors. Both experiments were repeated several times. The TPR and FPR values were calculated for different probabilities found using Eq. 2. As can be seen in Fig. 3a and b the TPR is maximum and FPR is minimum when the probability is 0.4 for the PIR sensor while the same is true at probability 0.7 for the AUS. Therefore, the thresholds τ_{PIR} and τ_{AUS} were set to 0.4 and 0.7 respectively for the detection experiments. Further, the priority of the PIR sensor was set higher than that of the AUS based on the TPR and FPR values attained.

To train the model 150 signals are captured and each signal is captured for 4 s. A signal is buffered for 4 s to ensure that buffering time of signal is greater than or equal to the processing time of signal. The data acquired as mentioned in Step 2 of Sect. 3 is only for the human being since the remaining non-human objects are countless and collecting data for this set is beyond the scope of this work. The data obtained from the PIR and AUS sensors was buffered separately and represented in form of matrices as shown by Eq. 1. Data was collected from each sensor for 600 seconds using only the human being as a target. Human beings dressed in different clothing formed targets for detection. Features, as mentioned in Step 2 in Sect. 3, were calculated for both the PIR and ultrasonic signals using the Eq. 3 through 10 below:

$$Maximum = Max\,(a_u)_{u=1}^V \tag{3}$$

$$Median = median\,(a_u)_{u=1}^V \tag{4}$$

$$Mean = \Sigma_{u=1}^V a_u \tag{5}$$

$$Root\ Mean\ Square = \sqrt{\frac{1}{V}\Sigma_{u=1}^V a_u} \tag{6}$$

$$Standard\ Deviation = \sqrt{\frac{1}{V-1}\Sigma_{u=1}V\,(a_u - Maximum)^2} \tag{7}$$

$$Kurtosis = V\frac{\Sigma_{u=1}^V\,(a_u - average)^4}{\left(\Sigma_{u=1}^V\,(a_u - average)^2\right)^2} \tag{8}$$

$$Crest\ Factor = \frac{\frac{1}{2}\,(Maximum - Minimum)}{RootMeanSquare} \tag{9}$$

$$Average\ Peak\ to\ Peak\ Distance(PPD) \tag{10}$$

where, V is size of the buffered signal and u_i is data instance within the buffered file.

In order to reduce the inherent noise in the raw signal, Discrete Wavelet Transformation (DWT) was performed on the AUS data before extracting its features. Figure 4 shows the nature of the ultrasonic signal reflected from a human being after applying DWT. As mentioned, the features were extracted from such transformed signals.

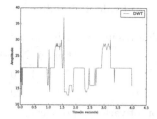

Fig. 4. Nature of the signal after DWT (Object: Human being)

Once the procedure defined in Step 2 of methodology section was completed, Step 3 was performed for different objects commonly present in an office and cafe environment.

At first features defined by Eq. 3 through 10 are extracted from the PIR sensor data. Using procedures detailed in Steps 4 through 6, the voting and probability of the confronted object are calculated on the basis of the PIR sensor signal. If this probability is found to be greater than τ_{PIR} then the object confronted is classified as a Human else as a Non-Human. If the object is classified as a Non-Human, the signal data received via the ultrasonic sensor for the same object is retrieved and its features are extracted after applying DWT. The associated probability is also calculated and voting performed accordingly. If the probability calculated for the AUS signal is less than τ_{AUS}, then the object is categorized to belong to the Non-Human class; else to the Human. To confirm the presence or absence of the human being, the platform on which the sensors were mounted was made to move to their left by a distance of 10 cm and then move back to the original position. During this motion, the new data is acquired via the sensors for the same obstacle present in front of the sensors. From this new data, the PIR sensor signal is once again analyzed in a similar way as defined earlier. If the new probability calculated for the PIR sensor is greater than τ_{PIR}, the presence of the human being is confirmed; else the object is deemed not belong to the Non-human class. Experiments were performed with different human beings, cushioned chairs, cupboards, wooden ply boards, cardboard, glass panels and cement and stone walls. Table 1(a), shows the results which indicates that out of 80 human beings, all were classified correctly with no false alarms(In column 1 of Table 1(a), the values in the brackets denote the number of obstacles confronted by the sensors). All non-human objects were also classified with high accuracy (100% in this case) calculated as defined by Eq. 11.

$$Accuracy = \frac{TPR + TNR}{TPR + TNR + FPR + FNR} \tag{11}$$

where, *TPR* is True Positive Rate, *TNR* is True Negative Rate, *FPR* is False Positive Rate and *FNR* is False Negative Rate. The results obtained using the voting based sensor fusion approach was compared with those obtained from Experiments 1 and 2 as well. When experiments are performed with the PIR sensor, it fails to detect a stationary human being (as expected) which caused a

Table 1. Results

(a) Classification results of Voting based sensor fusion

	Human	Non-human
Human (80)	80	0
Non-Human (120)	0	120

(b) Comparison of various approaches

	TPR	FPR
Voting-based sensor fusion	100	0
AUS($\tau_{AUS} = 0.7$)	80	5.71
PIR($\tau_{PIR} = 0.4$)	80	0

reduction in TPR. In case of the experiment with the AUS, only the cushioned chair and the glass panels were classified as human beings. However, these cases were correctly resolved when the voting based sensor fusion approach was used. Fusion allows detection of a moving human being using the PIR sensor signal and if stationary, the AUS performs this task. The double checking using AUS and the PIR sensor data obtained on movement of the robot back and forth provides for better and more reliable results. A comparison of the results obtained from the voting based sensor fusion with the results of classification obtained from the PIR as well as the AUS are shown in Table 1(b). The TPR and FPR values for the fusion approach were found to be 100 and 0 respectively. The same for the AUS and PIR at the predefined thresholds, were found to be 80 and 5.71 and 80 and 0 respectively. This indicates the superiority of the combined fusion based approach. Computations in the fusion based approach are performed only when the target is within a prescribed range. Further, if the PIR signal facilitates classification of the target as a human being, then the AUS signal is not processed. These aspects save on computational overheads.

4   Conclusions and Discussion

From the experiments performed to detect the presence of a human being using PIR, AUS and PING sensors one may conclude that the voting based sensor fusion approach performs far better than cases when individual sensors are used for the same task. The high TPR and low FPR values form the basis of this statement. Detection is not done unilaterally but is re-verified by other means using sensor data and also by physical movement relative to the target. The current approach does not however take into consideration the aspect of speeding objects. For such scenarios, the platform on which the sensors are mounted may need to match speeds in terms of mobility to bring down the relative speed and thus enable the detection. In future, we intend to augment this voting based fusion technique with more sensors. Such a sensor assembly mounted on a mobile robot will prove to be an efficient human detection system in hazardous environments as also scenarios where privacy is a major issue.

References

1. Akhoundi, M.A.A., Valavi, E.: Multi-sensor fuzzy data fusion using sensors with different characteristics. CoRR abs/1010.6096 (2010). http://arxiv.org/abs/1010.6096
2. Bai, Y.W., Cheng, C.C., Xie, Z.L.: Use of a time-variation ultrasonic signal and pir sensors to enhance the sensing reliability of an embedded surveillance system. In: 2013 26th Annual IEEE Canadian Conference on Electrical and Computer Engineering (CCECE), pp. 1–6, May 2013
3. Bai, Y.W., Cheng, C.C., Xie, Z.L.: Use of ultrasonic signal coding and pir sensors to enhance the sensing reliability of an embedded surveillance system. In: 2013 IEEE International Systems Conference (SysCon), pp. 287–291, April 2013
4. Can, G.N., akr, B., Naml, A.T., Dutaac, H.: Detection of humans from depth images. In: 2016 24th Signal Processing and Communication Application Conference (SIU), pp. 1477–1480, May 2016
5. Chen, Y., Rui, Y.: Real-time speaker tracking using particle filter sensor fusion. Proc. IEEE **92**(3), 485–494 (2004)
6. Damarla, R., Ufford, D.: Personnel detection using ground sensors (2007). http://dx.doi.org/10.1117/12.723212
7. Dong, B., Andrews, B.: Sensor-based occupancy behavioral pattern recognition for energy and comfort management in intelligent buildings. In: Proceedings of Building Simulation, pp. 1444–1451 (2009)
8. Feng, G., Liu, M., Guo, X., Zhang, J., Wang, G.: Genetic algorithm based optimal placement of pir sensor arrays for human localization. In: 2011 IEEE International Conference on Mechatronics and Automation, pp. 1080–1084, August 2011
9. Flores, S., Gei, J., Vossiek, M.: An ultrasonic sensor network for high-quality range-bearing-based indoor positioning. In: 2016 IEEE/ION Position, Location and Navigation Symposium (PLANS), pp. 572–576, April 2016
10. Gao, G., Zhang, Y., Zhu, Y., Duan, G.: Data fusion and multi-fault classification based on support vector machines. Vectors **3**(4), 5 (2006)
11. Georgescu, B., Shimshoni, I., Meer, P.: Mean shift based clustering in high dimensions: a texture classification example. In: Ninth IEEE International Conference on Computer Vision 2003, Proceedings, vol.1, pp. 456–463, October 2003
12. Gilmore, E.T., Frazier, P.D., Chouikha, M.: Improved human detection using image fusion. In: Proceedings of the IEEE ICRA 2009 Workshop on People Detection and Tracking, Kobe, Japan (2009)
13. Hondori, H.M., Khademi, M., Lopes, C.V.: Monitoring intake gestures using sensor fusion (microsoft kinect and inertial sensors) for smart home tele-rehab setting. In: 2012 1st Annual IEEE Healthcare Innovation Conference (2012)
14. Lim, M., Choi, J., Kim, D., Park, S.: A smart medication prompting system and context reasoning in home environments. In: Fourth International Conference on Networked Computing and Advanced Information Management, NCM 2008, vol. 1, pp. 115–118, September 2008
15. Machida, E., Cao, M., Murao, T., Hashimoto, H.: Human motion tracking of mobile robot with kinect 3D sensor. In: 2012 Proceedings of SICE Annual Conference (SICE), pp. 2207–2211, August 2012
16. Maslov, I.V., Gertner, I.: Multi-sensor fusion: an evolutionary algorithm approach. Inf. Fusion **7**(3), 304–330 (2006)
17. Moghavvemi, M., Seng, L.C.: Pyroelectric infrared sensor for intruder detection. In: 2004 IEEE Region 10 Conference TENCON 2004, vol. D, pp. 656–659, November 2004. (vol. 4)

18. Murphy, R.R.: Dempster-shafer theory for sensor fusion in autonomous mobile robots. IEEE Trans. Robot. Autom. **14**(2), 197–206 (1998)

19. Nadee, C., Chamnongthai, K.: Multi sensor system for automatic fall detection. In: 2015 Asia-Pacific Signal and Information Processing Association Annual Summit and Conference (APSIPA), pp. 930–933, December 2015

20. Nakamura, K., Zhao, H., Shibasaki, R., Shao, X.: Human sensing in crowd using laser scanners. INTECH Open Access Publisher (2012)

21. Park, J.T., Song, J.B.: Sensor fusion-based exploration in home environments using information, driving and localization gains. Appl. Soft Comput. **36**(C), 70–86 (2015). http://dx.doi.org/10.1016/j.asoc.2015.07.013

22. Politis, D.N., Romano, J.P.: Multivariate density estimation with general flat-top kernels of infinite order. J. Multivar. Anal. **68**(1), 1–25 (1999)

23. Sasiadek, J.Z.: Sensor fusion. Ann. Rev. Control **26**(2), 203–228 (2002)

24. Sidibe, Y., Druaux, F., Lefebvre, D., Leon, F., Maze, G.: Active fault diagnosis for immersed structure. In: 2013 International Conference on Control, Decision and Information Technologies (CoDIT), pp. 071–075, May 2013

25. Song, B., Choi, H., Lee, H.S.: Surveillance tracking system using passive infrared motion sensors in wireless sensor network. In: 2008 International Conference on Information Networking, pp. 1–5, January 2008

26. Sonia, T., A.M., Baruah, R.D., Nair, S.B.: Ultrasonic sensor-based human detector using one-class classifiers. In: 2015 IEEE International Conference on Evolving and Adaptive Intelligent Systems (EAIS), pp. 1–6, December 2015

27. Soyer, E.B.: Pyroelectric infrared (PIR) sensor based event detection. Ph.D. thesis, bIlkent university (2009)

28. Teixeira, T., Dublon, G., Savvides, A.: A survey of human-sensing: methods for detecting presence, count, location, track, and identity. ENALAB Technical report (2011)

29. Xu, Y., Wang, J.Y., Cao, B.X., Yang, J.: Multi sensors based ultrasonic human face identification: experiment and analysis. In: IEEE Conference on Multisensor Fusion and Integration for Intelligent Systems (MFI), pp. 257–261, September 2012

30. Yazar, A., Erden, F., Cetin, A.E.: Multi-sensor ambient assisted living system for fall detection. In: Proceedings of the IEEE International Conference on Acoustics, Speech, and Signal Processing (ICASSP14), pp. 1–3. Citeseer (2014)

31. Zigel, Y., Litvak, D., Gannot, I.: A method for automatic fall detection of elderly people using floor vibrations and sound - proof of concept on human mimicking doll falls. IEEE Trans. Biomed. Eng. **56**(12), 2858–2867 (2009)

Interface and Systems

A User Study About Security Practices of Less-Literate Smartphone Users

Pankaj Doke[1](✉), Sylvan Lobo[1], Anirudha Joshi[2], Nupur Aggarwal[2], Vivek Paul[2], Varun Mevada[2], and Abhijith KR[2]

[1] Tata Consultancy Services Limited, Mumbai, India
{pankaj.doke,sylvan.lobo}@tcs.com
[2] Indian Institute of Technology, Mumbai, India
{anirudha,nupuraggarwal,156330004,
156330007,156330003}@iitb.ac.in

Abstract. This paper describes the insights gained from user studies conducted with less-literate smartphone users in the context of Usable Smartphone Information Security. For the purpose of this study, we present the analysis and findings from 70 users. 37 users were selected using convenient sampling from a metropolitan city and 33 from a nearby town. We report findings on PINs and Pattern passwords, Phone as an Information Device and as a Commodity, Asset Valuation, Risk Identification, Risk Assessment, Risk Mitigation, Password Management, Privacy, amongst others.

Keywords: Usable security · Mobility · Smartphone · Low-Literacy · ICT4D · Passwords

1 Introduction

India has seen significant adoption of mobile phones and also has a significant number of citizens who are less-literate [1]. This opens up interesting research questions as users with low-literacy increasingly adopt and use smartphones in their daily lives. There is also a strong push towards mobile applications for various businesses. It seems users who have lower levels of literacy have weaker ability for abstraction and concepts [6, 7, 9, 10]. Consequently, when English language mediated smartphone adoption happens our users are at a significant disadvantage for their ability to leverage the systems for benefit, or ability to safeguard their interest since they are vulnerable to security and privacy cyber-mobile attacks using English.

In this paper, we are curious to understand Information Security related aspects on the axes of low-literacy and smartphone based devices and applications. We hope that our findings shed more light on how currently designed mobile phone security interventions are performing in the field.

© Springer International Publishing AG 2017
A. Basu et al. (Eds.): IHCI 2016, LNCS 10127, pp. 209–216, 2017.
DOI: 10.1007/978-3-319-52503-7_17

2 Literature Study

The ability to guess secretive information via either brute force or intelligent guesses has made information based security systems vulnerable. As a reactive response to the ever increasing threat of attacks from a computing system, mechanisms have been articulated to defeat or delay the compromise of secretive information. One of the fallouts of such designs has been that humans, for whom the systems have been designed, find it increasingly difficult to manage their security needs. If security systems were designed to be usable by humans then the compromises would be much lesser [11].

When it comes to less-literate users, information security can be a significant challenge owing to the cognitive load induced by mobile user interfaces in foreign languages. There exists a strong correlation between literacy and cognitive ability required to comprehend security aspects of a mobile interface. Ceci and Williams [10] indicate that variation in the amount of schooling is related to variation in the amount of intelligence. van Linden and Cremers [9] claim that functional illiterate users had low performance on tests which measured cognitive abilities and even on those which had no requirement of literacy. They also claim that such users have lower proficiency as compared to literate users even in the processing of spoken language. The authors also claim that functionally illiterate users have low visual organizational skills. The users perceive a complex figure as an aggregation than as a coherent entity. This has implications where the whole is larger than the sum. Typically, in interfaces which are to be consumed as a whole than as an aggregation of individual visual elements. For example, a window or a dialog box is an entity in its own right even though it is composed of smaller sub-units of visual elements with their own interaction mechanisms.

Reis and Castro-Caldas [8] indicate the relationship between literacy and visual representation of everyday objects and how less-literate users perform poorly. This indicates a possibility of flat interfaces or skeuomorphic interfaces performing differently than lesser and more literate users. Medhi et al. [5] investigated transfer of learning in the context of less-literate users and how the users performed poorly. In a study similar to ours, Chen et al. [3] discuss security perceptions of users in urban Ghana, a developing nation, however users were not less-literate and the focus was primarily on a computer.

In summary, prior literature as indicated above in cognitive psychology and its current context in ICTD, indicates a very strong correlation between literacy and performance of the users in an ICT context.

3 Method

The objective of this study was to glean insights into information security – practices, cognitive models, and processes adopted by less-literate users when using Android based smartphones. For a working definition of low-literacy, we chose 10 years as the maximum level of education. Overall 70 users were chosen for the study, of which 37 were from a metropolitan city and 33 from a town around 250 km away from the city. In the city, users were selected via convenient sampling, while in the town, the users were selected via an intermediary. Some users were also selected from a village adjoining the town. Our sampling strategy was door to door recruitment. Users were asked their education level and whether they owned a smartphone. There were 16 females and 54 males. The gender skew could be attributed to difficulty in arranging female smartphone users as participants, possibly due to the existing social aspects of way of life in rural India, where phones are generally shared in a family, with the male as the owner. The average age of the user was 36.83 years while the average education was 8.64 years.

The users were explained the nature of work and after seeking consent, the Contextual Inquiry method [2] was used to study the users. Field notes consisted of images of artifacts, users, observations and insights. The focus of the interviews were to assess the users perspectives with respect to asset valuation, risk perception, risk mitigation of the smartphone both as a tangible commodity and as an information device. The analysis was done using Affinity Diagrams based on guidance of the text 'Contextual Design' of Beyer and Holtzblatt [2].

4 Findings and Discussion

4.1 Commodity and Information

Field Note: *"Log sochenge ke ye koi toh hai. Pocket mein phone dalna is important"* (*"People will think, now this guy is someone important! It's important to keep the phone in the (shirt) pocket* – make it prominently visible, like a status symbol."*)

Field Note: *"She is very protective about her phone. She always changes the place of keeping the phone at her work place, being protective from clients, prospective clients and other people. She keeps the phone in her bag while travelling and removes it make/receive calls".*

We found that for users who treat the device as an Asset on the tangible dimension, the valuation is the direct mapping to their economic status and the Price which they paid for the purchase of the device. Such users tended to also treat the device as a status symbol and accordingly associate risks, namely arising due to loss of phone by theft or misplacement. On the other hand, some users also perceived and used the device actively as an Information Store for contacts in the phonebook, SMS and media content which are either user-generated or shared via social-media applications. The value transition happens when the user migrates from record keeping on paper (pocketbook) to digital tools (address book) on the phone. Here, the user is more likely to not only enforce physical security (not sharing with others) but also enforce Information Security. As the user moves towards more data centric applications like WhatsApp, there is a strong tendency to using Vault Applications or Password locks, essentially to protect the data and also maintain Privacy. The protection is observed when users tend to back up content on memory cards or in some cases contacts on the SIM cards. The user sentiment and behaviour is different towards the SIM card (Identity Management), memory cards (portable backup solutions) and phone (a housing mechanism for the SIM card, memory card and entertainment features). Thus, we observe a shift from the plane of commodity to the plane of an information device when the user migrates from tangible physical world objects to digital products (software) on the phone.

While the user is quickly able to respond to physical security, in the information plane, user responses are limited by their cognitive ability and awareness. To mitigate, they depend on the guidance by their peers. However, the user is still prone to misinformation as information shared by peers could be incorrect. There are popular misconceptions and notions about virus, e.g. to not switch on Bluetooth or not share media data as a virus could enter the phone. 'Virus' itself is a generic notion for device or software malfunction and instability. In the Information security space, users do not have conceptual models of the operations of the phone and the smartphone workings such as the operating system and user interface which leads to incoherency in responses.

4.2 Password Management

For users who identify with the smartphone as an Information Device, the protection is based on Information-based security. Accordingly, users either opt for PIN-based passwords or Pattern locks. In the following sub-sections, we cover two views *'Easy to recall, Easy to guess'* and *'Un-PIN me, I don't care'*.

4.2.1 "Easy to Recall, Easy to Guess"
Field Note: *"Gmail password is his phone number."*

Field Note: *"When she had a unlock password in her phone, a lot of people around her knew it since it read 2124, which looks like the name "Sh___" when written in Hindi, which is her husband's name. This was also the number of their Activa car."*

Due to the cognitive load associated with having to recall passwords or the implications, in terms of work impact or economic, when the password is forgotten, users tend to opt for patterns or PINs which are extremely easy to recall or guess. In this context, the recallability of the passwords has an overwhelming effect on the composition of passwords (very weak passwords). Given that the interface is in English language and the most familiar words in English language to them are their own names or those of their family members, these are chosen [8]. Unfortunately, users seem to believe that their passwords are strong or are completely oblivious to the fact that passwords need to be difficult to guess to others. While users do seem to choose passwords from their context, example initials of names, they could not be categorized as good choices.

Users also tend to map glyphs of the Devanagari character to glyphs in English or Roman and chose Patterns accordingly, when local language or cultural patterns themselves could be more secure. However, in spite of choosing easy to recall passwords, there were instances of users not being able to recall. In such situations instead of using recovery mechanism, they create new accounts, defeating Identity.

4.2.2 "Un-PIN Me, I Don't Care"
Field Note: *"I will just make new ID when I get a new phone."*

Field Note: *"He knows that there are multiple ways of recover if something goes wrong. But at the same time if one Google account password is forgotten he then makes a new account."*

In the Android ecosystem on mobile, the Gmail-id of the user serves as an Identity. In the event that they are unable to recall the password for the account, they create another account. They also seem to not use password recall mechanisms. So, from this perspective, the users really do not care for Identity Management on the phone, hence 'I don't care'. Such user behavior is likely to have impacts on the design of the mobile ecosystem. Apart from this, most of the triggers are when the users forget to recall their PINs or their patterns. In these situations, users try to disable PIN or Patterns Password features on their phone, hence the 'Un-Pin me'. From these aspects, one can draw inferences about user's temperament towards unable-to-recall situations. Instead of taking recourse to recovery mechanism, users tend to choose strong-handed choices of either not using the feature or circumventing the process. The users seem to be oblivious to the notion of digital identity and its associated aspects. On the other hand users are aware of passports or *Aadhar* (unique identity in India) which offer unique identity to them as they were perceived. While *Aadhar* offers a digital identity, it is operationally viewed as a paper identity. They are unable to apply the transfer learning model [5] to treat a Gmail account at the same level as an identity offered by a passport. Correspondingly, while they will protect the passport, they will be frivolous with a Google account creation since they are unable to accord the status of identity. Another aspect is the cost of creation for Gmail is zero.

4.3 Privacy

Users also seem to be aware about Privacy and exercise a certain amount of caution and precautionary steps when it comes to content they deem either objectionable (adult in nature) or that which their family members may have concerns about. Such privacy concerns however do not seem to be present when the users are in their peer circles. Social and cultural factors play a significant role as a deciding parameter to classify the content and behavioural response to the content handling. Accordingly, in the subsequent sections, we cover two categories, namely *'Forbidden Fruit'* where users lock only certain sections of the mobile applications and data. Most of these have their origins in the aspect of sharing of the phone in homes. Though one does observe that there is a tendency to have 'personal phones' which are 'out-of-bounds'. However, within the family, we expect the trend to continue, where phones of younger family members are used without permission or phones of elders are used by their children for gaming or 'rationed' usage of the internet - since 'internet' use is a privilege than a utility like electricity.

4.3.1 Forbidden Fruits
Field Note: *"Private cheese private rehna chahiye. (private stuff should stay private). People of my age send 'different kinds' of messages to each other, that cannot be made public. Others reading these may think why these (old) people use such language, send such pictures" – uses lock on WhatsApp".*

Field Note: *"Has a pin lock on WhatsApp, phone and contacts because they are "main things" in the phone. Wants to save it from her husband and the cleaning ladies who are there in the hostel. Put the lock two months ago by asking her daughter to do the same."*

Users tend to have some applications and certain content or storage areas which are 'off-limits' to everyone. These areas and applications are protected by PINs or Pattern locks. It is observed that when users are initiated into content sharing or social networking mobile applications, the initiator also coaches them about Privacy and locking mechanisms. Such users actively consume media via applications like WhatsApp.

4.4 Risk Mitigation

Risk Mitigators not only treat the phone as an information device of value, but, also go to great lengths to protect the information of value. Primarily, it has been observed that a contact in the phonebook and media is treated as an asset and valuation is proportional to the quantum of information. Users also exhibit risk enumeration from various events such as loss or theft of device, damage to device et cetera. Mitigation Techniques range from archiving the phonebook in their diaries or copying the contents onto another storage device.

Risk to the users can be discussed on two dimensions: (a) where the phone is a commodity, (b) where the phone is an information device. In the case of (a), the user treats the entity (phone) as any high value (compared to the income of the user) object. This is validated when the user's behaviour is modified when dealing with the phone. We observe such cases when the user deposits the phone in the safe of the drawer or in the cupboard/show-case, or keeps in personal close areas. In such cases the user resorts to physical security measures to secure the devices. In a reverse phenomenon we observe distress in the users when the phone is hidden or taken by known people. In the case of (b), users attached valuation in proportion to the volume of information on the device. In these cases, the measures to mitigate the risk are to backup to memory cards, computer, and physical diaries.

Users are particularly wary of their physical world reputation getting damaged due to security incidents. This is corroborated by actions of the users such as consciously clearing browser history, deleting images, messages and chat history, and using apps to clean the phone.

4.4.1 Save My Assets

Field Note – *"Phone update kiya toh sab gaane udd gaye"(I updated the phone, and lost all my songs")*.

Field Note: *"Keeps a backup of important numbers written down in a book back home (Mumbai). Because: 1 - he loses phones often. 2 - his father or elder brother might need to call someone up from home."*

Users who assess the smartphone as an information device also have simple models of valuation. Content such as phonebook, SMS and media are valued in proportion to their volume. A tendency to backup indicates either a temperament of risk mitigation or a feedback based learner due to undesirable events. Some users tend to archive their content, namely, contacts to diaries while some go to the higher levels of maintaining redundant or dedicated phones for phonebook or memory cards. The familiarity and convenience of a memory card also serves as a portable archive.

4.5 Discussion

To build a generalized conceptual model of how the phone and its security operates, and then to formulate an asset valuation and risk mitigation, thereby arriving at a response mechanism in line with the asset valuation using either a password or pattern-lock seems a ground for failure for less-literate users.

In respect of Medhi's work [5] we would like to argue that while the users were exposed to pattern locks which were alphabets of their name or the English alphabet there seemed to be no appropriation or modification to have a pattern which is nonlinguistic/non-English character. E.g. while *Rangoli's* are culturally rooted in India and traditionally taught with grid for patterns of geometric shapes, we did not find users merging the two concepts to do a geometric shape/rangoli pattern on the phone or even local language characters, in fact users would transliterate natively spoken words to English alphabet for the pattern. Within the devices, there exists an ability to increase the grid size. We find possible application of these findings in our less-literate user scenarios in terms of design interventions.

5 Conclusions and Future Work

Based on our findings we feel information security needs to be revisited in context of lesser literacy. The various modes of engagement which users adopt as well as the conceptual models they have, may not be the most appropriate mechanisms to deal with the threat posed by modern systems and attack mechanisms. The over-simplifications of response mechanisms or lower assessments of risks perceived, if at all, make users vulnerable. While the modern computational attack systems are less forgiving, users seem to be desirous of a benevolent world. Further work could be carried out to delve into how users model risks or can be taught security paradigms or concepts.

Acknowledgments. We thank all the users, volunteers, and all publications support and staff, who wrote and provided helpful comments on versions of this document.

References

1. DDW-0000C-08.xlsx.   http://www.censusindia.gov.in/2011census/C-series/C-08/DDW-0000C-08.xlsx. Accessed 19 Jul 2016
2. Beyer, H., Holtzblatt, K.: Contextual Design: Defining Customer-Centered Systems. Morgan Kaufmann Publishers Inc., San Francisco (2014)
3. Chen, J., Paik, M., McCabe, K.: Exploring internet security perceptions and practices in Urban Ghana. In: Symposium on Usable Privacy and Security (SOUPS), July 9–11, 2014, Menlo Park, CA (2014)
4. Egelman, S., Jain, S., Portnoff, R.S., Liao, K., Consolvo, S., Wagner, D.: Are you ready to lock? In: ACM SIGSAC Conference on Computer and Communications Security, pp. 750–761. ACM (2014)
5. Medhi, I., Menon, R.S., Cutrell, E., Toyama, K.: Correlation between limited education and transfer of learning. Inf. Technol. Int. Dev. **8**(2), 51 (2012)

6. Medhi, I., Menon, R.S., Cutrell, E., Toyama, K.: Beyond strict illiteracy: abstracted learning among less-literate users. In: ACM/IEEE International Conference on Information and Communication Technologies and Development (ICTD 2010), Article 23, p. 9. ACM, New York (2010). doi:http://dx.doi.org/10.1145/2369220.2369241
7. Medhi, I., Patnaik, S., Brunskill, E., Gautama, S.N.N., Thies, W., Toyama, K.: Designing mobile interfaces for novice and low-literacy users. ACM Trans. Comput. Hum. Interact. **18** (1), 28, Article 2 (2011). doi:http://dx.doi.org/10.1145/1959022.1959024
8. Reis, A., Castro-Caldas, A.: Illiteracy: a cause for biased cognitive development. J. Int. Neuropsychol. Soc. **3**(05), 444–450 (1997)
9. van Linden, S., Cremers, A.H.M.: Cognitive abilities of functionally illiterate persons relevant to ICT use. In: Miesenberger, K., Klaus, J., Zagler, W., Karshmer, A. (eds.) Computers Helping People with Special Needs, pp. 705–712. Springer, Heidelberg (2008)
10. Ceci, S.J., Williams, W.M.: Schooling, intelligence, and income. Am. Psychol. **52**(10), 1051 (1997)
11. Whitten, W., Tygar, J.D.: Why Johnny can't encrypt: a usability evaluation of PGP 5.0. In: 8th Conference on USENIX Security Symposium (SSYM 1999), vol. 8, p. 14. USENIX Association, Berkeley (1999)

Fuzzy Sliding Mode Controller for Dynamic Nonlinear Systems

Gaurav Kumawat[✉] and Jayashri Vajpai[✉]

Department of Electrical Engineering, M.B.M. Engineering College,
J. N. V. University, Jodhpur, Rajasthan, India
grv.kumawat@gmail.com, jvajpai@gmail.com

Abstract. Soft-computing based control is fast emerging as the method of implementation of intelligent automation in real world systems. Not only industrial systems, but a wide range of other diagnostic, predictive and autonomous control systems rely on soft computing tools, particularly on fuzzy logic, for the design of intelligent control system. This paper presents the design and application of a fuzzy sliding mode controller for dynamic nonlinear systems. The Takagi-Sugeno (T-S) fuzzy model has been used to represent the nonlinear system and design the fuzzy controller. The Parallel Distributed Compensation (PDC) has been used with sliding mode control to design the controller, with an aim to improve the performance of the system by output tracking with disturbance rejection. The proposed controller is better than the conventional PDC based fuzzy controller and has better output tracking efficiency robustness. The proposed methodology has been successfully employed for the well-established benchmark nonlinear system, viz., inverted pendulum on cart. Simulation results show that the designed fuzzy sliding mode controller tracks the desired output trajectory very fast, while rejecting disturbances.

Keywords: Fuzzy sliding mode controller · Takagi-Sugeno fuzzy model · Sliding mode control · Inverted pendulum

1 Introduction

Modern industrial world needs extensive interaction between plants, computers and humans. The autonomous control of the industrial equipment requires computer based controllers that emulate the performance of expert computer operators. Hence, Soft-computing based intelligent controllers are emerging as the preferred method of implementation of automation in a wide range of industrial systems. The real world systems that they cater to, are generally complex systems, with a large number of interacting variables, deeply ingrained uncertainties, sensitivity towards ambient noise and unmodelled dynamics, due to the presence of nonlinearities and time dependency of system parameters. Their controllers need several capabilities like tracking and predictive control, even in the presence of parameter variations, ambient noise signals and developing incipient faults. Hence, the controllers designed for the effective control and safe operation of these systems rely on soft computing tools, particularly on fuzzy logic, for the development of intelligent control system.

© Springer International Publishing AG 2017
A. Basu et al. (Eds.): IHCI 2016, LNCS 10127, pp. 217–228, 2017.
DOI: 10.1007/978-3-319-52503-7_18

Fuzzy logic control is an effective approach to design controllers for nonlinear systems, especially in the absence of complete knowledge of the plant, as in industrial processes. Fuzzy controllers are gaining popularity for advanced industrial control of complex and nonlinear systems, particularly those operating under the conditions of uncertainty and instability. A fuzzy controller has been designed by Yu et al. [1] to control nonlinear system with the minimum overshoot and settling time. ANFIS (Adaptive Neural-Fuzzy Inference System) has also been used by Hui et al. [2] to control nonlinear systems; this controller is found to be better than the conventional PID and LQR controllers. Fuzzy model based control is used widely because the design and analysis of overall fuzzy system can be performed by linear control theory. Sharma et al. [3], and Wang et al. [4] have designed a fuzzy controller based on Takagi-Sugeno (T-S) fuzzy model and parallel distributed compensation (PDC) to stabilize the non-linear systems.

In T-S fuzzy model, a nonlinear system is approximated by some linear sub-systems in the corresponding fuzzy regions of state space [5]. Thus, the antecedent parts contain fuzzy propositions but consequent parts contain a linear sub-system. The linear sub-systems are combined by fuzzy rules and the nonlinear system dynamics are expressed by weighted average of these linear sub-systems. The Parallel distributed compensation is used to design fuzzy controller [4]. A state feedback controller is designed for each linear sub-system of consequent part of T-S fuzzy model and finally are combined by fuzzy rules. These fuzzy controllers easily stabilize the nonlinear systems but their tracking efficiency is poor.

Sliding mode controller (SMC) is a robust controller with variable structure control that is widely used in industrial applications for stabilization and imparting robustness [6]. SMC uses a high speed switching control law to drive states of the nonlinear system onto a surface which is called sliding/switching surface and to maintain the states at this surface. The system behavior is modified according to switching function and the closed loop response becomes insensitive to nonlinearity. The parameter uncertainty, disturbance rejection and tracking problems are easily handled by SMC [7].

A fuzzy sliding mode controller has been designed in this paper for control of dynamic nonlinear systems. It is a combination of fuzzy controller with Parallel Distributed Compensation and sliding mode controller that improves the tracking efficiency of fuzzy controller by introducing a sliding control in it. The sliding mode controller acts as a supervisory controller over all the state feedback controllers. Such a controller has been proposed for uncertain discrete time systems by Mi et al. [8]. This paper presents a similar fuzzy sliding mode controller for continuous time nonlinear system. It is assumed that there is no parametric uncertainty in system, all states are measureable and noise is also present in system.

The paper is organized as follows: T-S fuzzy model and fuzzy controller is explained in Sect. 2; in Sect. 3 the fuzzy sliding mode controller is described, the proposed methodology is in Sect. 4 and in Sect. 5, T-S fuzzy model of inverted pendulum and fuzzy sliding mode controller is designed. The simulation results are shown in Sect. 6 and finally conclusion is drawn in Sect. 7.

2 Fuzzy Model Based Controller

The generalized dynamic nonlinear system is modeled by using Takagi Sugeno fuzzy model [1, 3]. This model is described by fuzzy IF-THEN rules, which represent local linear equivalent input-output relation of the system. The basic concept of T-S fuzzy model is to express the local dynamics of each fuzzy implication by a local linear system model.

2.1 Takagi-Sugeno (T-S) Fuzzy Model

The dynamics of nonlinear system is described by T-S fuzzy model with r rules and linear sub-systems in following format:

Rule i: IF $z_1(t)$ is M_{i1} AND.....AND $z_p(t)$ is M_{ip}

$$\text{THEN } \dot{x}(t) = A_i x(t) + B_i u(t)\, i = 1,2,\ldots,r \tag{1}$$

Where, r is total number of model rules, p is total number of premise variables, M_{ij} is fuzzy set for j^{th} premise variable of ith rule, $x(t) \in R^n$ is state vector, $u(t) \subset R^m$ is input vector, $A_i \in R^{n \times n}$, $B_i \in R^{n \times m}$. $z(t)$ is a vector containing all premise variables viz., $z_1(t)$, $z_2(t)$, ..., $z_p(t)$ i.e., $z(t) = [z_1(t)\ z_2(t) \ldots\ldots z_p(t)]$.

The final defuzzified output of the fuzzy system (3) is inferred as follows:

$$\dot{x}(t) = \frac{\sum\limits_{i=1}^{r} w_i(z(t))[A_i x(t) + B_i u(t)]}{\sum\limits_{i=1}^{r} w_i(z(t))} \tag{2}$$

Where, $w_i(z(t)) = \prod\limits_{j=1}^{p} M_{ij}(z_j(t))$, $w_i(z(t)) \geq 0\,\forall i$, $\sum\limits_{i=1}^{r} w_i(z(t)) > 0$

$w_i(z(t))$ is membership value of corresponding rule. $M_{ij}(z(t))$ is the grade of membership of $z_j(t)$ in M_{ij}.

By further simplification,

$$\dot{x}(t) = \sum\limits_{i=1}^{r} h_i(z(t))[A_i x(t) + B_i u(t)] \tag{3}$$

Where, $h_i(z(t)) = \dfrac{w_i(z(t))}{\sum\limits_{i=1}^{r} w_i(z(t))}$, $h_i(z(t)) \geq 0\,\forall i$, $\sum\limits_{i=1}^{r} h_i(z(t)) = 1$

Where, $h_i(z(t))$ is normalized grade of membership of corresponding rule. Model (3) describes the dynamics of T-S fuzzy model. If A_i, B_i for i = 1,2, ..., r are state controllable, then system is called locally state controllable, which is the necessary condition for designing of fuzzy controller.

2.2   Fuzzy Controller

The fuzzy controller (FC) is designed by Parallel Distributed Compensation (PDC). In PDC the local state feedback controller is designed for each linear sub-system of consequent of each rule of T-S fuzzy model. The FC is developed in similar fashion as described in Sect. 2.1 as follows.

The ith rule of Fuzzy controller is described as follows:

Rule i: IF $z_1(t)$ is M_{i1} AND. . . . . .AND $z_p(t)$ is M_{ip}

$$\text{THEN } u(t) = -K_i x(t) \, i = 1,2,\ldots,r \tag{4}$$

Where, r is number of controller rules and K_i is state feedback gain matrix for i^{th} linear sub-system of T-S fuzzy model. The final defuzzified output of the fuzzy controller is

$$u(t) = -\sum_{i=1}^{r} h_i(z(t))K_i x(t) \tag{5}$$

The overall closed-loop fuzzy system is obtained by substituting Eq. (5) into Eq. (3) as follows:

$$\dot{x}(t) = \sum_{i=1}^{r}\sum_{j=1}^{r} h_i(z(t))h_j(z(t))[A_i - B_i K_j]x(t) \tag{6}$$

A sufficient condition that guarantees the stability of closed loop fuzzy system is obtained on the basis of Lyapunov's direct method. The closed loop fuzzy system is globally stable, if there exists a common positive definite matrix P, such that:

$$(A_i - B_i K_i)^T P + P(A_i - B_i K_i) < 0$$

$$\left(\frac{(A_i - B_i K_j) + (A_i - B_j K_i)}{2}\right)^T P + P\left(\frac{(A_i - B_i K_j) + (A_i - B_j K_i)}{2}\right) \leq 0 \quad i < j \leq r \tag{7}$$

The stability of closed-loop fuzzy system depends on common positive definite P matrix that can be calculated by solving Linear Matrix Inequality (LMI) of Eq. (7). In the case where it is difficult to find this common positive definite P, the stability of closed-loop fuzzy system may not guaranteed. This fuzzy controller easily stabilizes the nonlinear systems but their tracking efficiency is very poor. The sliding controller easily handles the tracking performance while rejecting disturbance. Use of sliding mode control with fuzzy controller guarantees stability of fuzzy closed-loop system with improved robustness and tracking performance.

3   Fuzzy Sliding Mode Controller

In this section the fuzzy sliding mode controller (FSMC) construction is described. The FSMC is the combination of fuzzy controller by PDC and sliding mode controller. FSMC integrates the sliding mode control with local state feedback controllers of fuzzy controller. The sliding mode controller acts as a global controller over all local state feedback controllers. The FSMC capitalises on the advantages of both these controllers for robust tracking performance of overall system. The design of FSMC is as follows:

The state space model of a generalised dynamic nonlinear system with disturbance d(t), can be represented as:

$$\begin{cases} \dot{x}_1 = x_2 \\ \vdots \\ \dot{x}_{n-1} = x_n \\ \dot{x}_n = f(x(t)) + g(x(t))u + d(t) \end{cases} \qquad (8)$$

Where, $x(t) \in R^n$ is state vector, $u(t) \in R^m$ is input vector, $f(x(t))$ and $g(x(t))$ are known nonlinear function of system and $d(t)$ is disturbance present in system such as $|d(t)| \leq D$. The control objective is to obtain the state $x(t)$ for tracking a desired trajectory $x_d(t) = [x_{1d}(t), \ldots, x_{nd}(t)]$ in presence of disturbance $d(t)$.

The system given by Eq. (8) is represented by T-S Fuzzy model. The fuzzy sliding mode controller rules are just like fuzzy controller rules except that a sliding mode control input $u_{slide}(t)$ is also introduced. The ith rule for fuzzy sliding mode controller described as follows:

Rule i:IF $z_1(t)$ is M_{i1} AND......AND $z_p(t)$ is M_{ip}

$$\text{THEN } u(t) = -K_i x(t) + u_{slide}(t) \quad i = 1,2,\ldots,r \qquad (9)$$

The final defuzzified output of the fuzzy sliding controller is

$$u(t) = \frac{\sum_{i=1}^{r} w_i(z(t))[-K_i x(t) + u_{slide}(t)]}{\sum_{i=1}^{r} w_i(z(t))} = -\sum_{i=1}^{r} h_i(z(t))K_i x(t) + u_{slide}(t) \qquad (10)$$

The overall closed-loop fuzzy system is now obtained by substituting Eq. (10) into Eq. (8) as follows:

$$\dot{x}_n = f(x(t)) + g(x(t))\left[-\sum_{i=1}^{r} h_i(z(t))K_i x(t) + u_{slide}(t) \right] + d(t)$$

$$\dot{x}_n = f(x(t)) - g(x(t))\sum_{i=1}^{r} h_i(z(t))K_i x(t) + g(x(t))u_{slide}(t) + d(t)$$

Assuming,

$$F(x(t)) = f(x(t)) - g(x(t)) \sum_{i=1}^{r} h_i(z(t))K_i x(t) \tag{11}$$

Then,

$$\dot{x}_n = F(x(t)) + g(x(t))u_{slide}(t) + d(t) \tag{12}$$

The sliding surface is chosen as

$$s = e^{n-1} + a_1 e^{n-2} + \cdots + a_{n-1} e^1 \tag{13}$$

Where coefficients $a_1, \ldots, a_{n-1}$ are the coefficients of Hurwitz polynomial $s^n + a_1 s^{n-1} + \cdots + a_{n-1} s$, and, $e = x(t) - x_d(t) = [e^1 \ e^2 \ldots e^n]$ is the tracking error vector, with exponential reaching law defined as

$$\dot{s} = -\eta \, \text{sgn}(s) - ks$$

Differentiating the Eq. (13) with respect to time, gives:

$$\dot{s} = e^n + a_1 e^{n-1} + \cdots + a_{n-1} e^2$$

$$\dot{s} = \dot{x}_n - x_d^n + a_1 e^{n-1} + \cdots + a_{n-1} e^2 \tag{14}$$

On substituting the value of $\dot{x}_n$ from Eq. (12) to Eq. (14),

$$\dot{s} = F(x(t)) + g(x(t))u_{slide}(t) + d(t) - x_d^n + a_1 e^{n-1} + \cdots + a_{n-1} e^2 \tag{15}$$

Selecting the Lyapunov function

$$V = \frac{s^2}{2} \tag{16}$$

Differentiating the Eq. (16) with respect to time yields:

$$\dot{V} = s\dot{s} \tag{17}$$

On substituting the value of $\dot{s}$ from Eq. (15) to Eq. (17)

$$\dot{V} = s \left(F(x(t)) + g(x(t))u_{slide}(t) + d(t) - x_d^n + a_1 e^{n-1} + \cdots + a_{n-1} e^2 \right) \tag{18}$$

The control input of Fuzzy sliding mode controller is selected as

$$u_{slide}(t) = \frac{x_d^n(t) - F(x(t)) - \eta \, \text{sgn}(s) - ks - a_1 e^n - \cdots - a_{n-1} e^2}{g(x(t))} \tag{19}$$

Where $k > 0$, $\eta > 0$ are positive constants.
Substituting the value of $u_{slide}(t)$ from Eq. (19) in Eq. (18), gives:

$$\dot{V} = s(-\eta\,\mathrm{sgn}(s) - ks + d(t))$$

$$\dot{V} = -\eta|s| - ks^2 + sd$$

If η, $k \geq D$ where $|d(t)| \leq D$ then

$$\dot{V} = -\eta|s| - ks^2 + sd < 0$$

Therefore the closed-loop system (12) will be asymptotically stable if u(t) is used as input. So if η and k are selected properly, the disturbances are easily rejected with guaranteed fuzzy system stability.

If the T-S fuzzy model described in (3) is locally state controllable, then closed-loop fuzzy system described in (12) with control law (19) is asymptotically stable about the origin of coordinates.

The advantages of proposed controller are as follow:

1. The proposed controller improves tracking efficiency of fuzzy controller.
2. The overall stability of the system is guaranteed by sliding mode control hence there is no need to calculate common positive definite matrix P.

4 Proposed Methodology

The major steps in the design of the fuzzy sliding mode controller are as follows:

Step 1: Construct the T-S fuzzy model of dynamic nonlinear system.
Step 2: Construct the sliding mode controller by selecting sliding surface and control input as explained in Sect. 3.
Step 3: Select the desired poles according to the pre-specified desired dynamic response characteristics.
Step 4: For each local fuzzy sub-system, calculate the feedback gain.
Step 5: Obtain the fuzzy sliding mode controller by employing feedback gain, state vector and membership function of T-S fuzzy model and sliding mode control input.

5 Application

To illustrate this design approach, consider the problem of balancing an inverted pendulum on a cart which is a benchmark dynamic nonlinear system. The equations of motion of pendulum are [9]:

$$\dot{x}_1 = x_2$$
$$\dot{x}_2 = f(x(t)) + g(x(t))u + d(t) \qquad (20)$$

Where, $f(x(t)) = \dfrac{g\sin x_1 - aml\, x_2^2 \sin x_1 \cos x_1}{4l/3 - aml\cos^2 x_1}$, $\quad g(x(t)) = \dfrac{-a\cos x_1}{4l/3 - aml\cos^2 x_1}$

Where x_1 denotes the angle of pendulum (in radians) with respect to perpendicular axis and x_2 is the angular velocity of the pendulum, $g = 9.81$ m/s^2 gravity constant, m is the mass of pendulum, M is the mass of cart, 2l is the length of the pendulum, u is applied control force to the cart and $a = 1/(M + m)$. The simulation values are $M = 8$ kg, $m = 2$ kg, $l = 0.5$ m.

The proposed controller is designed to control the state $x_1(t)$ of inverted pendulum to track a desired trajectory $x_d(t) = 0.5\sin(\pi t)$ with disturbance $d(t) = 5\sin(2\pi t)$. The local approximation is used to construct T-S fuzzy model. The selected six operating points: $x_1 = [-\pi/3, 0, +\pi/3]$ & $x_2 = [-4, 0, +4]$ and the system (20) is linearised for nine possible combinations of these operating points. The following Fuzzy rules were obtained for T-S Fuzzy model as follows:

Model rules:

1. **IF** $x_1(t)$ is about $-\pi/3$ **AND** $x_2(t)$ is about -4 **THEN** $\dot{x}(t) = A_1 x(t) + B_1 u(t)$
2. **IF** $x_1(t)$ is about $-\pi/3$ **AND** $x_2(t)$ is about zero **THEN** $\dot{x}(t) = A_2 x(t) + B_2 u(t)$
3. **IF** $x_1(t)$ is about $-\pi/3$ **AND** $x_2(t)$ is about +4 **THEN** $\dot{x}(t) = A_3 x(t) + B_3 u(t)$
4. **IF** $x_1(t)$ is about zero **AND** $x_2(t)$ is about -4 **THEN** $\dot{x}(t) = A_4 x(t) + B_4 u(t)$
5. **IF** $x_1(t)$ is about zero **AND** $x_2(t)$ is about zero **THEN** $\dot{x}(t) = A_5 x(t) + B_5 u(t)$
6. **IF** $x_1(t)$ is about zero **AND** $x_2(t)$ is about +4 **THEN** $\dot{x}(t) = A_6 x(t) + B_6 u(t)$
7. **IF** $x_1(t)$ is about $+\pi/3$ **AND** $x_2(t)$ is about -4 **THEN** $\dot{x}(t) = A_7 x(t) + B_7 u(t)$
8. **IF** $x_1(t)$ is about $+\pi/3$ **AND** $x_2(t)$ is about zero **THEN** $\dot{x}(t) = A_8 x(t) + B_8 u(t)$
9. **IF** $x_1(t)$ is about $+\pi/3$ **AND** $x_2(t)$ is about +4 **THEN** $\dot{x}(t) = A_9 x(t) + B_9 u(t)$

Where,

$A_1 = A_9 = [0, 1; 12.64, 0.54]$, $A_2 = A_8 = [0, 1; 12.64, 0]$,
$A_3 = A_7 = [0, 1; 12.64, -0.54]$, $A_4 = A_6 = [0, 1; 14.47, 0]$,
$A_5 = [0, 1; 17.29, 0]$,
$B_1 = B_2 = B_3 = B_4 = B_7 = B_8 = B_9 = [0; -0.08]$, $B_5 = B_6 = [0; -0.17]$,

The membership functions for x_1, x_2 are chosen to be triangular membership functions, as shown in Fig. 1.

The sliding surface for the controller is selected as

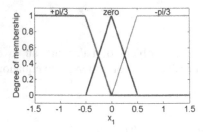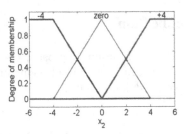

Fig. 1. Membership functions for state $x_1(t)$ and $x_2(t)$

$$s = 10e + \dot{e}$$

The magnitude of disturbance is $|d(t)| = 5$ and the pre-selected values $\eta = 15$ and $k = 10$. The control input of fuzzy sliding mode controller is, hence

$$u_{slide}(t) = \frac{\ddot{x}_d(t) - F(x(t)) - 15sat(s) - 10s - 10\dot{e}}{g(x(t))}$$

$sat(s(t))$ is used in place of $sgn(s(t))$ to avoid chattering. Here, $sat(s(t))$ is defined as follows

$$sat(s) = \begin{cases} +1 & s > +\delta \\ s/\delta & -\delta \le s \le +\delta \\ -1 & s < -\delta \end{cases}$$

δ is chosen to be 0.05 in this application. It is desired that the system response should have 10% maximum overshoot and 1 s settling time. Hence, the closed-loop poles locations are selected to be $s = [-4 + j5.454 \ -4-j5.454]$ for the system response to be under-damped with the designed transient specifications. The feedback gain matrixes are now calculated as follows:

$K_1 = K_9 = [-730.78, -106.88]$, $K_2 = K_8 = [-730.78, -100.12]$,
$K_3 = K_7 = [-730.78, -93.36]$, $K_4 = K_6 = [-341.17, -45.32]$,
$K_5 = [-357.17, -45.32]$,

The fuzzy rules of fuzzy sliding mode controller are now obtained as follows:
Controller rules:

1. **IF** $x_1(t)$ is about $-\pi/3$ **AND** $x_2(t)$ is about -4 **THEN** $u(t) = -K_1 \ x(t) + u_{slide}(t)$
2. **IF** $x_1(t)$ is about $-\pi/3$ **AND** $x_2(t)$ is about zero **THEN** $u(t) = -K_2 \ x(t) + u_{slide}(t)$
3. **IF** $x_1(t)$ is about $-\pi/3$ **AND** $x_2(t)$ is about +4 **THEN** $u(t) = -K_3 \ x(t) + u_{slide}(t)$
4. **IF** $x_1(t)$ is about zero **AND** $x_2(t)$ is about -4 **THEN** $u(t) = -K_4 \ x(t) + u_{slide}(t)$
5. **IF** $x_1(t)$ is about zero **AND** $x_2(t)$ is about zero **THEN** $u(t) = -K_5 \ x(t) + u_{slide}(t)$
6. **IF** $x_1(t)$ is about zero **AND** $x_2(t)$ is about +4 **THEN** $u(t) = -K_6 \ x(t) + u_{slide}(t)$
7. **IF** $x_1(t)$ is about $+\pi/3$ **AND** $x_2(t)$ is about -4 **THEN** $u(t) = -K_7 \ x(t) + u_{slide}(t)$
8. **IF** $x_1(t)$ is about $+\pi/3$ **AND** $x_2(t)$ is about zero **THEN** $u(t) = -K_8 \ x(t) + u_{slide}(t)$
9. **IF** $x_1(t)$ is about $+\pi/3$ **AND** $x_2(t)$ is about +4 **THEN** $u(t) = -K_9 \ x(t) + u_{slide}(t)$

The defuzzified output of the fuzzy sliding mode controller is

$$u(t) = -\sum_{i=1}^{9} h_i(z(t))K_i x(t) + u_{slide}(t)$$

6   Simulation Results

The performance of the proposed fuzzy sliding mode controller on the benchmark inverted pendulum on cart system was simulated for the following two cases;

- Natural response with non zero initial conditions.
- Forced response with nonzero initial conditions.

The results are compared with the fuzzy controller and are presented in following sub-sections:

6.1   Natural Response

For initial states $x_1(0) = 0.523$ rad, $x_2(0) = 0$ rad/sec, input is set to zero with disturbance $d(t) = 5\sin(2\pi t)$ the response of the closed-loop fuzzy system for the initial condition is shown in Fig. 2. Figure 2a shows the angular position of inverted pendulum and Fig. 2b. presents the control effort for both the proposed FSMC and FC.

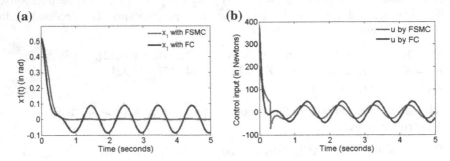

Fig. 2. Natural response for initial condition $x_1(0) = 0.523$ rad, $x_2(0) = 0$ rad/sec with disturbance $d(t) = 5\sin(2\pi t)$ (a) angular position $x_1(t)$ (2b) applied control effort $u(t)$

It is observed that the Fuzzy controller becomes inefficient when noise is introduced. The FSMC handles noise easily and all initial states sattle down to zero within 1 s.

6.2   Forced Response

The response of the closed-loop fuzzy system for initial states $x_1(0) = 0.523$ rad, $x_2(0) = 0$ rad/sec, and input for desired trajectory $x_d(t) = 0.5\sin(\pi t)$ with disturbance d (t) = $5\sin(2\pi t)$ is shown in Fig. 3. Figure 3a shows the angular position of inverted pendulum, and Fig. 3b. compares the control effort of the two controllers.

The results show that the FSMC is much more capable to efficiently track the desired state trajectory than the FC. Further it can be seen that the error is always present in the response of the FC, whereas it reduces to zero in the case of FSMC. The FSMC tracks the desired state trajectory even in presence of noise within 1 s.

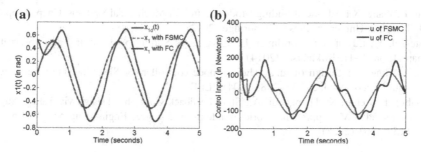

Fig. 3. Forced response for initial condition $x_1(0) = 0.523$ rad, $x_2(0) = 0$ rad/sec with disturbance $d(t) = 5\sin(2\pi t)$ (a) angular position $x_1(t)$ (b) applied control effort $u(t)$

7 Conclusion

Soft computing based controllers are gaining popularity in the computer based control of complex, real world industrial systems, as they are capable of performing at the same level as human experts. This has been made possible largely due to the inclusion of fuzzy computing in the well-established control technology. The methodology of implementing fuzzy sliding mode controller for the dynamic nonlinear systems, by using state feedback, has been proposed and developed in this paper. The proposed controller has been designed and implemented for the benchmark inverted pendulum on cart system and its performance has been compared with the fuzzy controller. The simulation results show that the proposed controller improves the tracking performance of the closed loop fuzzy control system. Further, it has been shown that it has good capabilities of rejecting noise. The overall stability of nonlinear system is guaranteed by the incorporation of sliding mode control. The authors are presently working for extending the design of proposed controller for the systems in which the states are not available for feedback, by implementing an observer based controller. Further, the FSMC can be designed with nonlinear sliding surface to remove the reaching phase of sliding mode controller.

References

1. Yu, L.H., Jian, F.: An inverted pendulum fuzzy controller design and Simulation. In: 2014 International Symposium on Computer, Consumer and Control, pp. 557–559 (2014)
2. Hui, Y.X., Sheng, L.H., Ping, L.G., Fu, X.G.: Control experiment of the inverted pendulum using adaptive neural - fuzzy controller. In: 2010 International Conference on Electrical and Control Engineering, pp. 629–633. IEEE (2010)
3. Sharma, B., Awasthy, N.: State feedback controller design via T-S fuzzy model. In: Sixth International Conference on Fuzzy Systems and Knowledge Discovery, pp. 176–181 (2009)
4. Wang, H.O., Tanaka, K., Griffin, M.F.: An approach to fuzzy control of nonlinear system: stability and design issues, pp. 14–23. IEEE (1996)
5. Tanaka, K., Wang, H.O.: Fuzzy Control System Design and Analysis- A Linear matrix inequality Approach, Chap. 3, pp. 5–81. Wiley, New York (2001)

6. Liu, J., Wang, X.: Advanced Sliding Mode Control for Mechanical Systems, Chap. 2, pp. 41–80. Tsinghua University Press, Cambridge (2012)
7. Shtessel, Y., Edwards, C., Fridman, L., Levant, A.: Sliding mode control and observation, Chap. 1, pp. 1–41. Birkhäuser (2014)
8. Mi, Y., Jing, Y.: Design of fuzzy sliding mode controller for SISO discrete-time systems. J. Control Theo. Appl. **3**, 253–258 (2004). IEEE
9. Khaber, F., Zehar, K., Hamzaoui, A.: State Feedback Controller Design via Takagi-Sugeno Fuzzy Model: LMI Approach, World Academy of Science, Engineering and Technology, pp. 1158–1163 (2008)

Enhancing Usability Inspection Through Data-Mining Techniques: An Automated Approach for Detecting Usability Problem Patterns of Academic Websites

Kalpna Sagar[(⊠)] and Anju Saha

University School of Information and Communication Technology,
GGSIPU, Sector 16-C, Delhi 110078, India
sagarkalpna87@gmail.com

Abstract. Usability is one of important attribute of software quality. It is associated with the "ease of use" of any system. Usability evaluation is becoming significant component of software development. Usability evaluation is performed through qualitative assessments. Qualitative assessments can be attained through Qualitative usability inspection (QUI). QUI methods emphasize on evaluating the interface of a specific system. These methods turn out to be complicated when huge number of systems related to similar context of use, are considered jointly to impart a general diagnosis. The principal cause for this is due to substantial quantity of information that should be conceptualized simultaneously. To handle substantial quantity of information, this paper proposes a novel approach that integrates QUI with automated woorank tool and data-mining techniques (association rules and decision tree). To validate this proposed approach, 50-academic websites are evaluated and usability problems patterns related to academic websites are identified by processing 2475 records.

Keywords: Usability · Usability engineering · Qualitative usability inspection · Heuristic evaluation · Context of use · Usability problem patterns · Data-mining knowledge discovering in databases · Association rules · Decision trees

1 Introduction

Usability is one of most important software quality attribute, as highlighted in various standards and usability models, e.g. ISO 9241-11 [1], 9126 [2], McCall [3] and Boehm [4] etc. It is associated with the "ease of use" of a given software system [S5]. ISO 9241-11 defines usability as "the extent to which a product can be used by specified users to achieve specified goals with effectiveness, satisfaction and efficiency in a specified context of use" [1]. "Context of use" means the description of actual conditions under which an interactive system is being evaluated. Context analysis [6, 7] is required for conducting research on software usability.

Usability evaluation is generally done through qualitative and quantitative assessments. Qualitative assessments are performed by evaluation team during evaluation

© Springer International Publishing AG 2017
A. Basu et al. (Eds.): IHCI 2016, LNCS 10127, pp. 229–247, 2017.
DOI: 10.1007/978-3-319-52503-7_19

stage. This stage is very crucial because it empowers developers to encompass expert feedback till admissible level of usability is achieved. Generally, research practitioners focus on qualitative assessments of usability. Quantitative assessments summarize the results in single metrics after evaluating various dimensions of software usability [8, 9]. Also, Quantitative assessments are not strong enough to compute overall usability of a software system [8, 10]. This motivates us to focus on qualitative assessments of usability. So, an attempt has been made to achieve qualitative results through Qualitative usability inspection (QUI).

Some models are defined for usability evaluation of interactive systems [11–13]. Usability evaluation has also been done on early prototypes with data-mining techniques [14, 15]. The context of use as a whole has been addressed but only homepage of academic websites are considered for usability evaluation. Further, there are several existing data-mining approaches for assessing and evaluating the usability [17–20]. Furthermore, usability evaluation has been done in different domains e.g. mobile [21–24] and website [25–29]. Usability evaluation has been significantly done for web-domain [26]. Hence, it also motivates us for usability evaluation in web-domain, mainly for academic websites.

Nowadays, one of the major challenges for usability evaluation is identifying the common usability related problems for a context of use as a whole. Recognising such problems can assist novice research practitioners in evaluating an advanced interface belonging to same context and to restrain errors when a new interactive system is being produced. To overcome with this challenge, QUI is employed that emphasizes on the "what" over the "how many" questions belonging to identification of usability related problems for a context of use as a whole.

QUI provides information about usability problems that occurs due to violations of heuristics belonging to a system. Identified usability problems belonging to a system, are prioritized. And, then a usability document is generated, containing a prioritized list of identified usability problems. QUI incorporates various divergent methods that emphasize on evaluating the interfaces of the systems [5, 8, 10]. These methods turn out to be complicated when substantial number of systems related to same context of use, are considered jointly to impart an extensive diagnosis. The principal cause for this is due to substantial quantity of information that should be conceptualized and handled simultaneously.

To handle substantial quantity of information, this paper proposes a novel approach called QUI_c[1]. This qualitative usability inspection is based upon data-mining techniques. Association rules and decision trees are used to find the common usability related problems for a context of use as a whole [30]. Each identified usability problem pattern belongs to a relevant usability characteristic of academic websites. Such usability problem patterns are then analysed through data-mining knowledge discovery process. Then, the usability document comprises the final output that is a list of prioritized usability problems.

---

[1] QUI_c represents Qualitative usability inspection for context of use.

To validate this proposed approach, an experiment is conducted in which academic websites[2] are evaluated and common usability problems patterns related to these websites are identified. In other words, a general diagnosis for the context of use of academic websites is analysed by employing QUI_c approach. The experiment involves processing of 2475 records that store qualitative information. The evaluation team uses heuristic evaluation and defines the 55-heuristic questions related to academic websites. Each heuristic question represents usability features of academic websites. Each feature is related to usability attribute of academic websites (shown in Table 1)[3]. For usability evaluation of websites, data is collected manually that is time consuming and cumbersome task [15]. So, evaluation team decides to use automated tool that provides answers to maximum defined heuristic related questions within minimum resources of

Table 1. Correspondence between heuristic-related questions belonging to Mobile (8 questions), Social Media (6 questions) and Security (3 questions) Category with attributes name

Attribute name	Heuristic questions	Possible answers
Mobile category		
Font size legibility	Is website's text readable on mobile devices?	*No:* When web page's text is not legible (i.e. may be too small) on mobile devices. *Yes:* When web page's text is legible on mobile devices. *NI(Needs Improvement)*: When web page's text needs improvement for font size legibility.
Mobile compatibility	Does website require any plugin or embedded object to load on mobile?	*No:* When website contains embedded objects. *Yes:* When website looks perfect because it does not contain any embedded objects like Flash, Silverlight or Java so that content can be accessed. *NI:* When website needs improvement for mobile compatibility.
Mobile frameworks	Does website use mobile frameworks to load perfectly in multiple devices?	*No:* Mobile frameworks have not been detected. *Yes:* Mobile frameworks have been detected. *NI:* When website needs improvement so that mobile frameworks can't be detected.

(continued)

---

[2] Top-50 academic websites listed in National Institutional Ranking Framework are considered.

[3] The detailed description of attribute along with categories are listed in Table 1.

Table 1. (*continued*)

Attribute name	Heuristic questions	Possible answers
Mobile friendliness	Is website optimized for users on mobile browsers?	*No:* When website is poorly optimized for visitors on mobile Visitors *Yes:* When website is super optimized for visitors on mobile devices. *NI*: When website is fairly optimized for mobile visitors.
Touchscreen readiness	Are website's menu/links/buttons are perfectly large enough to be easily readable and tapped on mobile devices?	*No:* When website does not have most important buttons/links to be large enough to be tapped easily. *Yes:* When website has the most important buttons/links perfectly large enough (atleast 48 pixels in height and width) to be tapped easily. *NI:* When website needs improvement for buttons/links.
Mobile rendering	Does website render nicely with all the features on mobile which user sees on desktop?	*No:* When website is not rendered nice on mobile device. *Yes:* When website is rendered nice on popular mobile devices. *NI*: When website needs improvement for mobile rendering.
Mobile speed	Does website load in mobile device with high speed?	*No:* Mobile speed finds to be slow, when website runs on mobile. *Yes:* Mobile speed finds to be fast, when website runs on mobile device. *NI:* Mobile speed finds to be average, when website runs on mobile.
Mobile viewport	Does website contain well configured viewport so that content fits within the specified viewport size?	*No:* When the website does not specify a viewport or viewport is not well configured. Or the content does not fit within the specified viewport size. *Yes:* When website contains well-configured viewport and the content is within the specified viewport size. *NI:* When website has mobile viewport but content does not fit within the specified viewport size (i.e. needs improvement).

(*continued*)

Table 1. (*continued*)

Attribute name	Heuristic questions	Possible answers
Social Media Category Blog	Does website contain a blog to engage user and to increase online visibility?	*No:* When a Blog is not found on the website. *Yes:* When a Blog is found on the website. *NI:* When website needs improvement for Blog.
Facebook page	Is university's website socially active on social networking sites i.e. Facebook as more than 5 million traffic come from social media?	*No:* When website/university does not have its Facebook page. *Yes:* When website/university have its Facebook page. *NI:* When website/university has its Facebook page but needs improvement.
Google+ page	Does university website contain Google + page?	*No:* When website/university does not have its Google + page. *Yes:* When website/university have its Google + page. *NI:* When website/university has its Google + page but needs improvement.
Google+ publisher	Does website provide Google + publisher tag to socialize their pages on the social network?	*No:* When website is missing a rel="Publisher" tag for linking to Google + Page. *Yes:* When website has rel="Publisher" tag for linking to Google + Page. *NI:* When website needs some improvement.
Twitter account	Is university's website socially active on social networking sites i.e. twitter?	*No:* When account is available but not registered. *Yes:* When website/university have its Twitter account which is booked and linked to website. *NI:* When website/university has its Twitter account which is booked but not linked to website.
Related websites	Does website contain any other related websites links/URLs to get information about how other competitors are doing?	*No:* When related websites are not found. *Yes:* When other related websites (some may be competitors while other may be websites with related contents) are found on website (max of 4 websites). *NI:* When website needs improvement for related websites

(*continued*)

Table 1. (*continued*)

Attribute name	Heuristic questions	Possible answers
Security		
Server signature	Is university website's server signature feature disabled which are highly recommended for security reasons?	*No:* When server signature is enabled on website. *Yes:* When server signature is disabled on website for security of website. *NI:* When website can be slightly improved for server signature.
Robots.txt	Does website have Robot.txt file to prevent search engines robots from accessing specific directories and pages?	*No:* When Robots.txt file is not found on website. *Yes:* When website has Robots.txt file. *NI:* When website needs improvement for Robots.txt file which may contain error.
SSL secure	Is university website's using SSL certificate to have secure transaction or encrypted connection between users and website's server?	*No:* When website is not SSL secured *Yes:* When website is SSL secured so that confidential information can be protected between user and server. *NI:* SSL secure feature needs improvement.

time and money. In this study, these answers are collected with help of automated tool named woorank [31].

The remaining part of this paper is organised as follows: Sect. 2 presents related work. Section 3 gives an overview of major concepts of data-mining knowledge discovery from data repository, association rules and decision trees. Section 4 provides details of proposed approach that augment the conventional QUI process with woorank tool and data-mining knowledge discovery process. Such novel approach identifies common usability problems patterns for context of use as a whole, but mainly from a qualitative viewpoint. Section 5 reports experimental results that validate the proposed approach. Section 6 summarizes conclusion and future work. This paper is motivated by some preliminary research work done on usability evaluation of early prototypes through association rules [14], and usability testing through datamining techniques [15, 16].

2 Related Work

Various studies have discussed the usability inspection methods [42, 43]. Some articles have explained the state of art of methodologies and the qualitative usability evaluation [8, 44–47]. For usability evaluation of interactive systems, some models are defined [11–13]. Usability evaluation of early prototypes with data-mining techniques e.g. association rule, is explained in [14]. Few studies have integrated usability evaluation with KDD process (association rules and decision trees) for finding significant patterns belonging to academic websites [15, 16]. The context of use as a whole has been addressed but only homepage of websites are considered. [17] uses association rule mining for assessing the usability of system. [18] provides recommendation models that are explained as a set of association rules to enhance the usability of the system. Even, the decision trees are applied for improving the usability evaluation [19, 20]. But these techniques are not used to analyse usability of a context of use as a whole. Software usability evaluation has also been done using fuzzy approaches [48, 49] and model driven development approach [28]. Furthermore, usability evaluation has been done in different domains e.g. mobile [21–24] and website [25–29].

3 An Overview of Data-Mining Knowledge Discovery from Data Repository

Data-Mining Knowledge Discovery from data repository is an approach that extracts previously unknown and potentially constructive information from the available data. It is a process that normalizes the data from heterogeneous sources and stores it in data repository or database. From data repository, attributes are generally identified. This extraction process in data-mining knowledge discovery [32] is executed as follows:

1. Data pre-processing: It consists of a number of steps that are prerequisite to produce data for mining process, e.g. (a) Data cleaning: The aim of this step is to remove irrelevant data and fill in missing values etc. (b) Data integration: Data from heterogeneous sources may be combined into single repository or database. (c) Data Selection: Relevant data required for evaluation, are retrieved from the database (d) Data transformation: Data are transformed into appropriate forms for mining process.

2. Data-mining: After pre-processing the data, various data-mining techniques can be executed on data, to extract significant patterns from it. There are several data-mining techniques available in literature e.g. Association rules, Classification, Clustering, Prediction, Sequential Patterns, Decision trees, trend analysis, Bayesian data analysis, Regression, neural network models etc. [33]

3. Pattern Identification: After applying datamining algorithms, patterns are identified that represents undiscovered knowledge. These patterns are then evaluated by using different measures.

4. Graphical representation: Different visualization and presentation techniques are employed to represent mined knowledge or results to users.

A software platform for Data-Mining Knowledge Discovery process generally involves 3-components (See Fig. 1.): (1) **Data Repository** to store the data (2) a **Data-Mining Knowledge Discovery engine** for conducting and executing datamining operations. This engine offers various datamining algorithms and modules for pattern evaluation and visualization (3) a **Query Interface** that assists user to interact with engine by posing queries in datamining query language. It is important to note that datamining is most vital step in Data-Mining Knowledge Discovery process, since it extracts previously unknown information and identifies hidden patterns of information for evaluation of patterns.

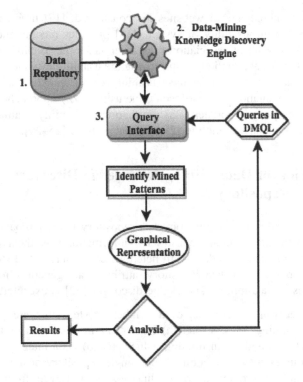

Fig. 1. Major components of data-mining knowledge discovery process platform

Data mining tasks demands the use of robust software platforms. As the number of existing platforms are persistently growing and therefore, choosing the suitable platform is becoming complicated task. Further, Different free-open source platforms have also been used for conducting data-mining knowledge discovery process [34]. e.g. RapidMiner [35], R [36], Orange [37], and Weka [38, 39]. Each platform has their own particular DMQL language that varies from basic command-line interpreters to sophisticated tools.

This paper is based upon some preliminary research work done by Gonzalez et al. [14], on usability evaluation of early prototypes through association rules and usability testing through data-mining techniques [15, 16]. For usability evaluation, Weka is generally used for applying data-mining techniques. Hence, the tool used in this study is Weka [39].

Furthermore, in this paper, we present an extended approach to the conventional qualitative usability inspection process by integrating 2-datamining techniques, e.g. association rules and decision trees. In next sub-section, we summarize main characteristics of these 2-techniques.

3.1   Association Rules

Association rules help in identifying significant relationships among attributes in a given data set [30]. The identification of these relationships is generally used in decision making processes for real world problems, e.g. market-basket analysis etc. In other words, association rules guide us to identify various patterns in datasets. For creating association rules, various algorithms have been used for mining knowledge from huge databases, e.g. **Apriori,** Predictive Apriori, Filtered Associator and FPGrowth [30]. The thresholds values are generally specified for confidence and support values so that computational complexities can be reduced.

3.2   Decision Trees

A decision tree is a flow-chart-like tree structure, where each node denotes a test on an attribute value, each branch represents an outcome of the test, and tree leaves represent classes or class distributions. In order to classify an unknown sample, the attribute values of the sample are tested against the decision tree [33, 38]. Therefore, the decision trees can be used in predicting the behavior of context of use under evaluation. Furthermore, decision trees can be generated from relatively small transactional databases to identify "target attribute". Various algorithms have been used for creating decision tree e.g. ID3, J48, FT and C4.5 [33]. During construction of decision trees, 66.6% data is used for training whereas remaining 33.3% is used for testing.

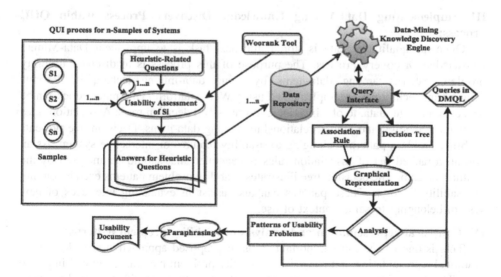

Fig. 2. Proposed approach for Qualitative Usability Inspection process (QUI_c) for interactive systems.

4   A Proposed Approach for Usability Evaluation

In this section, a novel approach of QUI_c process is proposed that is based upon the integration of a conventional usability inspection [40, 41], an automated woorank tool [31] and Data-Mining Knowledge Discovery Process [32]. Figure 2. represents architecture of the proposed approach. The proposed approach of QUI_c process is composed of the following steps:

I. Formulating QUI_c process:
   The QUI_c process starts with the selection of team-members for evaluation team. The evaluation team chooses a sample of interactive systems, $W_c = \{W1...Wn\}$ that is taken to present entire context C. Further, the team selects appropriate method for usability evaluation of system. The automated tools required for usability evaluation are also being explored. Furthermore, this step considers the selection of particular software that is required for implementation of Data-Mining Knowledge Discovery Process.

II. Defining the Heuristic related questions and gathering QUI_c data:
   Heuristic evaluation is a frequently used method for usability evaluation [42, 26]. The evaluation team defines heuristics-related questions. Each question belongs to various usability features or attributes of interactive systems that are to be evaluated. The correspondence between each heuristic question with attributes is shown in Table 1. The QUI_c process is based upon qualitative data that is collected using automated tool. The collected data is stored in data repository e.g. database and spreadsheet etc.

III. Implementing Data-Mining Knowledge Discovery Process within QUI_c process:
   Once the qualitative data is stored, the next task is to implement Data-Mining Knowledge Discovery Process. The purpose of this process is to discover usability problem patterns among data-items by using data-mining techniques. Various data-mining techniques are implemented using Weka platform. Different queries can be posed to mine the data, to obtain association rules and decision trees. Association rules represent unknown and hidden relationships among data items. Decision tree presents usability problem patterns belonging to usability features of interactive system. As an output, a ranked list of association rules is generated that should be analysed by the evaluation team. The decision tree illustrates usability problem patterns each belonging to usability features. These patterns can also assist in evaluating interfaces of new system belonging to same context of use.

IV. Graphical representation and analysis of usability problem patterns:
   This is one of the most important steps of proposed approach. As, it deals with graphical representation and analysis of usability problem patterns generated in above step. The evaluation team uses different visualization tools provided by Weka. These tools help the team in visualising different graphics belonging to generated usability problem patterns. For components in ranked lists of rules, results are shown as a sequence of association rules (as shown in Table 3.). The decision trees are very descriptive by themselves in their tree like representation (as shown in Fig. 3). Further,

evaluation team analyses each obtained usability problem patterns to detect usability problems belonging to interactive system of same context of use. Every detected usability pattern is discussed on the basis of its significance (e.g. support and confidence). The relative rank of each association rule with respect to entire ranked list (as shown in Table 3) is related to prioritization of the usability problem patterns. For admissible decision trees, the evaluation team decides to consider the threshold values of 70% for classified instances. These admissible decision trees can be added in final usability document.

V. Paraphrasing and reporting QUI$_c$ conclusions in usability document:

The QUI$_c$ process generates the output as a set of general usability problems patterns. Each detected usability problem pattern should be paraphrased as a usability problem in standard format [5] as shown in Table 3. If needed, then different visualizations of the usability problem patterns should be attached to final output using this format.

5 Experimentation and Results

In this section, obtained results are provided after applying proposed approach on 50-academic websites.

Step 1: First step is to select the evaluation team-members. The evaluation team is formed of 2 members: 1-usability experts, 1-Associate professor in university, 1-PHD student. All members of team have knowledge about usability and are frequent users of academic websites. The Google chrome browser is used to visualize the academic websites. Ideally, QUI$_c$ process must consider comprehensive usability evaluation of all academic websites in context of use. But practically, it is not a feasible task because of huge associated resources in terms of time and cost. Therefore, a sample of academic websites W$_c$ = {W1...Wn} is taken to present entire context C. The guidelines have been followed to impart a representative sample for academic websites [22]. As a sample, websites of **top 50-academic universities** of India listed in National Institutional Ranking Framework[4] are considered. Furthermore, the evaluation team searches for automated tool for usability evaluation of academic websites. And then, decides to use automated tool, named woorank [31]. This tool provides answers to maximum defined heuristic related questions within minimum time. The output spreadsheet contains answers of heuristic related questions of the website evaluation. Moreover, the planning step considers the selection of particular software that is required for implementation of Data-Mining Knowledge Discovery Process. So, Weka is used for applying data-mining techniques e.g. association rules and decision trees.

Step 2: The evaluation team has chosen heuristic evaluation as it is frequently used methods for usability evaluation [10, 42, 26]. The team defines 55-general heuristic-related questions. Each question belongs to various characteristic of academic websites that are to be evaluated. Range values for each possible answer are also defined.

---

[4] https://www.nirfindia.org/univ.

Each heuristic-related question is associated with an attribute name of website usability that woorank tool evaluates[5]. Table 1. represents correspondence between heuristic-related questions belonging to Mobile, Social Media and Security category with associated attribute name and possible answers. Due to the space constraint, only three categories that represent latest technological issues are defined and presented in Table 1.

Woorank tool is employed to evaluate an academic website as a whole[6]. During evaluation of each website, tool generates pdf file. This file contains answer for each heuristic question in terms of green, red and orange color. Green color represents that particular attribute or usability feature is available (or gets pass) on that website and we assign it as "yes" in output spreadsheet. Similarly, red color represents that particular attribute or usability feature is not available (or contains error) on that website and we assign it as "no" in output spreadsheet. Orange color represents that particular attribute or usability feature is available on that website but needs improvement and we assign it as "ni" in output spreadsheet. In this way, answers of maximum defined heuristic questions are evaluated and collected. These answers are stored in output spreadsheet that represents usability data. Now, each heuristic question is linked with an attribute in output spreadsheet. The attribute's names appear as a column heading in output spreadsheet. Each row of output spreadsheet corresponds to complete usability evaluation of a specific academic website. Hence, 45 websites are evaluated and 55 heuristic questions are defined and their answers are stored into output spreadsheet, resulting into 2475 records.

Step 3: After collecting and storing qualitative data, the next task in QUI_c process is to implement Data-Mining Knowledge Discovery Process. The aim of this step is to discover usability problem pattern among attributes by using data-mining techniques. Evaluation team uses weka software platform and query interface to pose queries for data-mining techniques. Association rules and decision tree are implemented on qualitative data. The association rules express unknown and hidden relationships among attribute representing usability data. For implementing association rules, apriori algorithm is used. As an output, ranked list of rules is obtained and is presented in Table 2. The evaluation team uses this list to identify the hidden relationship among attributes. The decision tree is also implemented that presents usability problem pattern belonging to usability features of academic websites. Further, these patterns can assist in evaluating new websites belonging to same context of use. For implementing decision tree, ID3 algorithm is used. As an output, decision tree is generated and presented in Fig. 3. The evaluation team considers decision tree representation as new knowledge that helps in identifying usability problems pattern concerning the context of use of academic websites as a whole.

Step 4: WEKA supports apriori algorithm that generates a ranked list of best 11-association rules (with support 70% and confidence 90%). Table 2 represents

[5] In this approach, 7-different categories are considered, namely Design, Content, Navigation, Security, Search, Mobile, and Social Media, for evaluation of usability of academic websites.

[6] Out of 50-top universities, woorank could not evaluate 5-universities due to some security reasons.

Table 2. Ranked list L1 of association rules belonging to heuristic-related questions (generated by WEKA)

Ranked list of Best 11-Association Rules	
#1	Mobile_Rendering = Yes 43 ==> Mobile_Frameworks = No 43 conf:(1)
#2	Mobile_Compatibility = Yes 35 ==> Mobile_Frameworks = No 35 conf:(1)
#3	Mobile_Speed = No 33 ==> Mobile_Frameworks = No 33 conf:(1)
#4	Mobile_Rendering = Yes Mobile_Compatibility = Yes 33 ==> Mobile_Frameworks = No 33 conf:(1)
#5	Mobile_Rendering = Yes Mobile_Speed = No 31 ==> Mobile_Frameworks = No 31 conf:(1)
#6	Mobile_Compatibility = Yes 35 ==> Mobile_Rendering = Yes 33 conf:(0.94)
#7	Mobile_Compatibility = Yes Mobile_Frameworks = No 35 ==> Mobile_Rendering = Yes 33 conf:(0.94)
#8	Mobile_Speed = No 33 ==> Mobile_Rendering = Yes 31 conf:(0.94)
#9	Robots_txt = Yes SSL_Secure = NI 7 ==> Server_Signature = No 7 conf:(1)
#10	Google+_Page = No 40 ==> Google+_Publisher = No 40 conf:(1)
#11	Related_websites = Yes Google+_Page = No 39 ==> Google+_Publisher = No 39 conf:(1)

* Association rules that describe similar usability problems are collectively considered.

obtained ranked list L1 of association rules belonging to heuristic related questions of **Mobile, Security and Social Category.**

It is important to note that association rules from L1 list impart initial guidelines to identify problematic attributes in discovered patterns. These problematic attributes generally represent usability problems. Therefore, association rules with problematic attributes are only considered. For example, association rules **#1, #2, #3, #4, and #5** in L1 show that there exist usability problem patterns related with absence of Mobile_Frameworks in academic websites. Another example, association rules **#10 and #11** in L1 show that there exist usability problem patterns related with absence of Google+_Publisher in academic websites. The usability problem patterns also help in detecting significant relationships among attributes. By just looking at output spreadsheet, the evaluation team cannot easily detect these significant relationships that are found in association rules **#1, #2, #3, #4, and #5** in L1. The discovered patterns can impart constructive information for evaluating the usability of academic websites that is not considered in National Institutional Ranking Framework but relates to same context of use. Further, these usability problem patterns emphasize on significant heuristic-related questions that should be carefully considered during usability evaluation. The usability problem patterns can be considered as guidelines for development of novel academic webpages.

The Evaluation team can pose some more queries to further mine the information form output spreadsheet. But, queries must have different support and confidence values. A query can be used to determine some other problematic attributes in L1 (mention in following example).

```
Mobile_Viewport = No
| Touchscreen_Readiness = NI
| | Mobile_Speed = No
| | | Mobile_Compatibility = NI: No
| | | Mobile_Compatibility = Yes: NI
| | | Mobile_Compatibility = NF: null
| | | Mobile_Compatibility = No: null
| | Mobile_Speed = NI: No
| | Mobile_Speed = NF: null
| | Mobile_Speed = Yes: null
| Touchscreen_Readiness = No
| | Mobile_Compatibility = NI: No
| | Mobile_Compatibility = Yes
| | | Mobile_Speed = No: No
| | | Mobile_Speed = NI: No
| | | Mobile_Speed = NF: null
| | | Mobile_Speed = Yes: No
| | Mobile_Compatibility = NF: null
| | Mobile_Compatibility = No: No
| Touchscreen_Readiness = Yes: null
| Touchscreen_Readiness = NF: null
Mobile_Viewport = Yes
| Mobile_Rendering = Yes: Yes
| Mobile_Rendering = NF
| | Touchscreen_Readiness = NI: null
| | Touchscreen_Readiness = No: No
| | Touchscreen_Readiness = Yes: Yes
| | Touchscreen_Readiness = NF: null
Mobile_Viewport = NF: NF
Mobile_Viewport = NI
| Touchscreen_Readiness = NI: Yes
| Touchscreen_Readiness = No: Yes
| Touchscreen_Readiness = Yes: NI
| Touchscreen_Readiness = NF: null
```

Fig. 3. Decision tree by using ID3 algorithm (target attribute Mobile_friendliness with possible values {Yes, No, NI, NF}) (visualization provided by WEKA platform)

GetRules(Output_Spreadsheet) where
[Antecedent has {(Mobile_Frameworks = sometimes)}and support >0.8 and confidence >0.9]}

The evaluation team poses more queries to get decision trees. The decision trees generally help in predicting the behavior of context of use under evaluation. WEKA selects 66.66% of available data from the output spreadsheet as training set and remaining data as test set. For admissible decision trees, the threshold values are set to 70% of classified instances. Figure 3. provides textual representation of resulting decision tree for any target attribute e.g. Mobile_friendliness.

Every branch of decision tree can be read as "if-then" rule. Figure 3 represents rule as: if (Mobile_Viewport = No) and (Touchscreen_Readiness = NI) and (Mobile_-Speed = No) and (Mobile_Compatibility = NI) then (Mobile_friendliness = No). It is important to notice that values stored in the attributes Mobile_Viewport, Touchscreen_Readiness and Mobile_Speed are significant to determine the value of attribute Mobile_friendliness (as they are adjacent to root of the decision tree). The value stored in the Mobile_Compatibility is not extremely relevant for predicting the value for Mobile_friendliness. This stored information in decision tree is previously unknown and hidden for evaluation team. Hence, evaluation team considers decision tree representation as new knowledge that helps in identifying usability problems concerning the context of use.

Step 5: The QUI$_c$ process generates the output as a set of usability problem patterns. It is recommended to paraphrase each pattern as a usability problem. Each usability problem should be reported and documented in a standard format [5] given in Table 3.

Table 3. Usability document reporting usability problem

Category	Usability problem pattern	Frequency (1 to 5)	Justification	Evaluation team comment	Recommendations
Mobile	**#1** The absence of Mobile_Frameworks is related with poor speed of mobile devices.	2	Association rules #3 and #5.	The usability problem pattern identifies problem in academic websites related to speed on mobile devices.	Recommend a mobile framework to ensure high speed on mobile devices.
Security	**#2** The Server_Signature feature is enabled that relates with non-secure (SSL) connection.	1	Association rule #9.	Security issues are recognized in usability problems patterns of websites.	Must Disable the server signature feature on the website to ensure secure SSL.
Social	**#3** The absence of Google+_Publisher is associated with absence of Google +_Page.	2	Association rule #10 and #11.	Official Google + Page is not identified on academic websites.	Include Google "Publisher" tag for linking to Google+ Page to socialize the network.

6 Conclusions and Future Work

QUI methods emphasize on evaluating the interfaces of the systems. These methods turn out to be complicated when substantial number of systems related to same context of use, are considered jointly. To overcome with this problem, a novel approach of QUI_c process is proposed that is based upon the integration of a conventional usability inspection (HE), an automated woorank tool and Data-Mining Knowledge Discovery Process. The study provides usability problem patterns highlighting the latest technological issues in academic websites. Identifying such problem patterns and proposing solution to these problems can help the developers and usability professional in various aspects. The insight about these patterns can assist developers to restrain from these problems, when new academic website is being developed. Further, detecting such problems at early stages can reduce development efforts in time and costs. Various instincts that were informally expressed by evaluation team during QUI_c process can now be significantly examined through data-mining knowledge discovery process. Therefore, for real world problems (i.e. usability problem patterns in Indian academic websites), the study can conclude that proposed integrated approach, can be executed successfully. The primary advantage of proposed approach is detection of unknown and hidden relationships among attributes belonging to academic websites. Another advantage is detection of usability problems patterns in academic websites. In comparison to previous related works, this study addresses certain differences as follows:

1. To the best of author's knowledge, there is no alternative approach that extends traditional QUI process with automated woorank tool and data-mining knowledge discovery process, as presented in this paper.
2. Qualitative usability inspection like HE method performs usability evaluation that is closer to real user insights. And thereafter, generates more relevant results.
3. In proposed approach, an attempt has been made to consider latest technological issues related to academic websites (e.g. mobile, social media and security etc.). 55 Heuristics related questions are defined for evaluating the usability of academic websites. The study presents only 17 heuristic questions due to space constraint.
4. In this study, 7 categories are considered for defining heuristic questions related to academic websites. These categories are defined to identify the usability problem patterns belonging to latest technological issues in mobile domain etc. Whereas other study has defined only 4 categories [15] without considering these issues.
5. Other study [15] has evaluated only homepages of academic websites. In this study, the automated woorank tool is used to provide extensive evaluation of entire academic websites. Using this tool, the data is collected within limited time period as well.
6. The proposed approach highlights the fact that academic websites, generally, are not mobile optimised. To best of author's knowledge, this issue is not addressed in any study.
7. Another issue is related with server signature feature that found to be enabled on some websites. These full server signatures can be exploited and attacked by hackers. Hence, academic websites are not fully secured.

8. Although, Google+ has reached 100 million users faster than Facebook and Twitter. But academic websites do not have official Google+ Page to socialize them on the social network. The proposed approach highlights another fact that academic websites do not have Google+ _Publisher for linking to Google+ _Page.

The future work can be done in employing different algorithms for association rules and decision trees. Such advanced algorithms can provide more extensive results. One can also emphasize on testing different ranking functions for association rules. Further, the generated data is compiled into databases that can be used for statistical analysis. Such analysis can help in quantitative assessment of usability for academic websites. In near future, research can be pursued in these directions.

References

1. International Organization for Standardization. ISO 9241-11: Ergonomic requirements for office work with visual display terminals (VDTs). Part 11: Guidance on usability. Geneva, Switzerland (1998)
2. ISO 9126: Information Technology Software Product Evaluation Quality Characteristics and Guidelines for their Use, Geneva (1991)
3. McCall, A.J., Richards, K.P., Walters, F.G.: Factors in Software Quality, vol. II. Rome Aid Defence Centre, Italy (1977)
4. Boëhm, B.: Characteristics of Software Quality. TRW Series on Software Technology, vol. 1. North-Holland, Amsterdam (1978)
5. Dumas, S.J., Redish, C.J.: A Practical Guide to Usability Testing. Intl. Specialized Book Service Inc, Portland (2000)
6. Maguire, M.: Context of use within usability activities. Int. J. Hum. Comput. Stud. **55**, 453–483 (2001)
7. Bevan, N., Kirakowsi, J., Maissel, J.: What is usability? In: Proceedings of the 4th International Conference on HCI, pp. 651–655 (1991)
8. Ivory, Y.M., Hearst, A.M.: The state of the art in automating usability evaluation of user interfaces. ACM Comput. Surv. **33**, 470–516 (2001)
9. Sauro, J., Kindlund, E.: A method to standardize usability metrics into a single score. ACM HCI (2005) [S12]
10. Nielsen, J.: Usability Engineering. Morgan Kaufmann, San Francisco (1993)
11. Beyer, H., Holtzblatt, K.: Contextual Design: Defining Customer-Centered Systems. Morgan Kaufmann, San Francisco (1998)
12. Constantine, L., Lockwood, L.: Software for Use: A Practical Guide to the Models and Methods of Usage-Centered Design. Addison-Wesley, Reading (1999)
13. Mayhew, J.D.: The Usability Engineering Lifecycle: A Practioner's Handbook for User Interface Design. Morgan Kaufmann, San Francisco (1999)
14. González, M.P., Granollers, T., Lorés, J.: A hybrid approach for modelling early prototype evaluation under user-centred design through association rules. In: Doherty, G., Blandford, A. (eds.) DSV-IS 2006. LNCS, vol. 4323, pp. 213–219. Springer, Heidelberg (2007). doi:10.1007/978-3-540-69554-7_17
15. Gonzalez, M.P., Lores, J., Granollers, A.: Enhanching usability testing through datamining techniques: a novel approach to detecting usability problem patterns for a context of use. Inf. Softw. Technol. **50**, 547–568 (2008)

16. Gonzalez, M.P., Granollers, T., Lores, J.: Métricas predictivas de la usabilidad: un nuevo enfoque para su ponderación cualitativa (available in Spanish). In: Proceedings of the VI Spanish Conference on Human Computer Interaction, pp. 233–241 (2005)
17. Tiedtke, T., Martin, C., Gerth, N.: Awusa: a tool for automated website usability analysis. In: 9th International Workshop DSVIS (2002)
18. Alipio, J., Pocas, J., Azevedo, P.: Recommendation with association rules: a web mining application. In: Data Mining and Warehouses Conference IS-2002 (2002)
19. Finlay, J.: Machine learning: a tool to support improved usability. Appl. Artif. Intell. **11**, 633–665 (1997)
20. Sikorski, M.: Beyond product usability: user satisfaction and quality management. In: CHI 2000 Extended Abstracts Human Factors Computing Systems, pp. 61–62. ACM Press (2000)
21. Moumane, K., Idri, A., Abran, A.: Usability evaluation of mobile applications using ISO 9241 and ISO 25062 standards. SpringerPlus. **5**, 1–15 (2016)
22. Jankowski, J., Grabowski, A.: Usability evaluation of vr interface for mobile robot teleoperation. Int. J. Hum. Comput. Interact. **31**, 882–889 (2015)
23. Ji, G.Y., Park, H.J., Lee, C.: A usability checklist for the usability evaluation of mobile phone user interface. Int. J. Hum. Comput. Interact. **3**(3), 207–231 (2006)
24. Kjeldskov, J., Stage, J.: New techniques for usability evaluation of mobile systems. Int. J. Hum. Comput. Stud. **60**, 599–620 (2004). Elsevier
25. Erickson, W., Trerise, S., Lee, C.: The accessibility and usability of college websites: is your website presenting barriers to potential students? Commun. Coll. J. Res. Pract. **37**, 864–876 (2013)
26. Fernandez, A., Insfran, E., Abrahão, S.: Usability evaluation methods for the web: a systematic mapping study. Inform. Softw. Technol. **53**, 789–817 (2011). Elsevier
27. Aziz, M.A., Isa, W.A.R.W.M., Nordin, N.: Accessing the accessibility and usability of malaysia higher education website. In: International Conference on User Science and Engineering, pp. 203–208. IEEE (2010)
28. Panach, I.J., Juristo, N., Valverde, F.: A framework to identify primitives that represent usability within model-driven development methods. Inform. Softw. Technol. **58**, 338–354 (2015). Elsevier
29. Lee, S., Cho, J.E.: Usability evaluation of Korean e-Government portal. In: Stephanidis, C. (ed.) UAHCI 2007. LNCS, vol. 4556, pp. 64–72. Springer, Heidelberg (2007). doi:10.1007/978-3-540-73283-9_8
30. Han, J., Kamber, M.: Data Mining: Concepts and Techniques. Morgan Kaufmann, San Francisco (2000)
31. Woorank. https://www.woorank.com/
32. Pyle, D.: Data Preparation for Data Mining. Morgan Kaufmann, San Francisco (1999)
33. Mitchell, T.: Machine Learning. McGraw Hill, New York (1997)
34. Mikut, R., Reischl, M.: Data mining tools: wiley interdisciplinary reviews. Data Min. Knowl. Disc. **1**, 431–443 (2011)
35. https://rapidminer.com/
36. http://www.rdatamining.com/
37. Demsar, J., Zupan, B., Leban, G.: Orange: From Experimental Machine Learning to Interactive Data Mining White Paper. Faculty of Computer and Information Science, University of Ljubljana (2004)
38. Witten, I.H., Frank, E.: Data Mining: Practical Machine Learning Tools and Techniques. Morgan Kaufmann, San Francisco (2005)
39. WEKA. http://www.cs.waikato.ac.nz/ml/weka/

40. Mack, R., Nielsen, J.: Usability inspection methods. ACM SIGCHI Bull. **25**(1), 28–33 (1993)
41. Rivero, L., Barreto, R., Conte, T.: Characterizing usability inspection methods through the analysis of a systematic mapping study extension. Clei Electron. J. **16**(1), 11 (2013)
42. Hollingsed, T., Novick, G.D.: Usability Inspection Methods after 15 Years of Research and Practice (2007)
43. Wharton, C., Rieman, J., Lewis, C., Polson, P.: Usability Inspection Methods: The Cognitive Walkthrough Method: A Practitioners Guide. Wiley, New York (1994)
44. Hornbæk, K.: Current practice in measuring usability: challenges to usability studies and research. Int. J. Hum. Comput. Stud. **64**(2), 79–102 (2006)
45. Paterno, F.: Model-Based Design and Evaluation of Interactive Application. Springer, New York (2000)
46. Rosson, M., Carroll, J.: Usability Engineering: Scenario-Based Development of HCI. Morgan Kaufmann, San Francisco (2002)
47. Sauro, J., Kindlund, E.: A method to standardize usability metrics into a single score. In: ACM HCI (2005)
48. Dubey, K.S., Gulati, A., Rana, A.: Usability evaluation of software systems using fuzzy multi-criteria approach. Int. J. Comput. Sci. Issues **9**(3), 404 (2012). ISSN: 1694-0814
49. Dubey, K.S., Rana, A., Sharma, A.: Usability evaluation of object oriented software system using fuzzy logic approach. Int. J. Comput. Appl. **43**(19), 35–41 (2012). ISSN: 0975-8887

Personalized E-library: A Recommender System Based on Learner's Feedback Model

Rajni Jindal and Alka Singhal[(⊠)]

Department of Computer Science, Delhi Technological University, Delhi, India
alkasinghal83@gmail.com

Abstract. The paper proposes an E-library which will be combining the collective intelligence of the social web with E-learning. In this paper, we argue that though it is important to have a structured explicit knowledge base in an E-learning site, in the present scenario when lot of research is being done on the personalized E-learning, it is necessary to refine the repository with the tacit knowledge which is unstructured and implicitly resides in people. It can be utilized from socialization, feedbacks, reviews, blogs etc. The paper suggests an e-library system where a derived structured knowledge base will be generated from both explicit and tacit knowledge of the learners and in return will give learner a targeted, updated and quality material to refer. In this paper, we have designed an E-library which will be user oriented and will estimate the utility of every resource for a given user based on two metrics Level and Like. It also provides a feedback system for modifying the metrics with the user response. It considers the effect of expert biasing and also diversity of collective intelligence.

Keywords: E-learning · E-library · Collective intelligence · Wisdom of crowd · Personalized E-learning · User centric systems

1 Introduction

Tom Gruber says, "The Social Web is an ecosystem of participation, where value is created by the aggregation of many individual user contributions" [1]. Whereas E-learning is to provide the right knowledge to the right person at the right time using internet technologies [2]. If we explore Social Web in E-learning to create a Social E-learning ecosystem, it will be a knowledge system made by the people and for the people. This will generate a platform for learning which will be user centric, updated, just in time and interactive. The paper suggests an e-library system where a derived structured knowledge base will be generated from both explicit and tacit knowledge of the learners and this will give in return learner a targeted, updated and quality material to refer.

The organization of the paper is as follows: Sect. 2 defines the factors which became the reason for studying the relevant work and choosing the field for research, Sect. 3 gives basic conceptual idea about the wisdom of crowd which will be acting as the basic pillar for the framework defined in the later section for the e-library. Section 4 gives a brief survey of the published papers in the field of collective intelligence and E-learning. It also gives a graphical notation of the publishing trend in the field of Collective intelligence alone and with E-learning. Section 5 defines the basic

© Springer International Publishing AG 2017
A. Basu et al. (Eds.): IHCI 2016, LNCS 10127, pp. 248–263, 2017.
DOI: 10.1007/978-3-319-52503-7_20

architecture of the proposed application including its components and working. Section 6 describes various case studies to judge the performance of the feedback system. And finally concluding with Sect. 7, it concludes with the benefits and challenges faced by the proposed application.

2   Motivation

The success of an E-learning system resides in five characteristics [2]: Personalization, Interactive, Just in time, Current and User centric. It can be explained as "E-learning system should provide information after analyzing the learner's objectives and her skill level with the courses knowledge base assembled exactly on what learner needs to know". It should provides "give and take" learning, not mere textbook static learning. Knowledge is complex and multifaceted to be captured within a learner model. The amount of knowledge that can be captured represents only the tip of the iceberg. The bulk of the iceberg below the waterline represents the knowledge in learner's heads, i.e. their tacit knowledge (Polanyi, 1967).

A major impediment in the online E-libraries is that the search is provided based on the keywords. The authenticity of the content is within the boundaries established by the subject matter experts. Learner's pace and expertise is not majorly taken into account while providing the search results in E-library. There is a wide disconnect between learner's needs and choice of material delivered to her on a given relevant topic.

For a helpful and successful E-library, it is required that the search results provided to the learner should be in accordance with her expertise and requirements. And also the knowledge base should be updated at a regular basis on learner's feedback and the learner should be provided with all the details like level, readability and other basic information of the subject material with the search result.

It can be well understood that the two main bases for all successful E-libraries are collaboration and user centric application. Collaborative learning provides interaction and a regular updation whereas user centric makes it personalized and focused to the learner.

The paper also follows the 3P Learning model defined in [14] and provides Personalization, Participation and knowledge Pull in context of E-library described in the later section. This paper harness the implicit and explicit knowledge of the Learner to improve the quality and provide a user centric search result in E-library.

3   Vision of Wisdom of Crowd

In today's world, Web Users are no longer hesitant to express themselves. They want to share their experience with products and services, they have used and want to help the other users by their feedback. This can be through posting interesting article, rating a product or service they bought or used, or generating new content [3]. Now users are not confined to follow and use the same products or services, used by their older generation. They can take help of the feedback and review to select the best among the available resources. This increased user interaction and participation helps to generate tacit knowledge also known as Wisdom of Crowd that can be converted into intelligence in the web applications.

Surowiecki, (2004) mentions four conditions that characterize wise crowds: (a) Diversity of opinion (each person should have some private information, even if it is just an eccentric interpretation of the known facts), (b) Independence (people's opinions are not determined by the opinions of those around them), (c) Decentralization (people are able to specialize and draw on local knowledge), and (d) Aggregation (some mechanism exists for turning private judgments into a collective decision).

The core concept of Collective intelligence of users [3]

- The intelligence is generated by the users themselves through collective set of interactions and contributions made by them.
- This intelligence further acts as a refiner for what's valuable in the application for a user. This collective intelligence aids in relevant searching and making decisions to the users.

4 Literature Review

The success of E-learning exploits collective intelligence to achieve user centric learning. This section presents a survey of the volume of the publications published in relevance with development of collective intelligence in general (graph I) (from 2000 to 2015) and generically development of collective intelligence in the field of E-learning (graph II) (from 2000 to 2015). The details are obtained from Google Scholar, March 2016. The graphs shows a tremendous growth in the volume of publications in the specified stream of research from 2000 to 2010 as this is the phase of origin of Web 2.0 and later which inspired E-learning to evolve into E-learning 2.0 (Figs. 1 and 2).

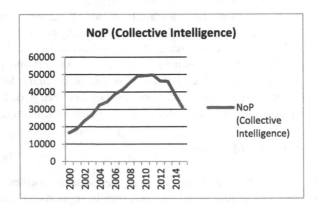

Fig. 1. Number of publications (NoP) in field of collective intelligence

After the volume of publications, the section presents a literature survey of the major contributions in the field of collective intelligence and E-learning from year 2005 to 2015. Table 1 shows These all publication uses different theories to facilitate Personalized e-learning under the concept of collective intelligence.

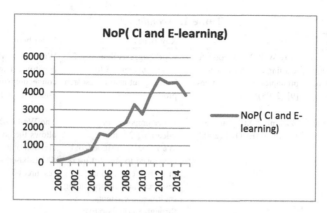

Fig. 2. NoP in field of collective intelligence with E-learning

Table 1. Survey of the major contributions in the field of Personalized E-learning from year 2005 to 2015

Author	Title	Key issues	Contribution
Ilhami Görgün, Ali Türker, Yildiray Ozan, and Jürgen Heller.	Learner modeling to facilitate personalized E-LEARNING experience [4] (2005)	Describes learner modeling strategy in an adaptive learning system to provide learner with personalized e-learning experience.	Utilize ontological abstraction (OWL) for the learner modeling
Chih-Ming Chen, Hahn-Ming Lee, Ya-Hui Chen	Personalized e-learning system using item response theory [5] (2005)	In personalized systems, learner ability should also be considered with learner's preferences, interests and browsing behaviors in implementing personalization mechanism.	Utilize IRT which consider both course material difficulty and learner's ability to provide individual learning path. It also use MLE, explicit learner feedback and collaborative voting for precise estimation
Chao Boon Teo, Robert Kheng Leng Gay [6]	A knowledge-driven model to personalize E-Learning [6] (2006)	It argues that most of the learning system are just being online without providing personalization.	It provides methodology for eliciting and personalized tacit knowledge using concept maps
Weihong Huang, David Webster, Dawn Wood and Tanko Ishaya	An intelligent semantic e-learning framework using context-aware Semantic Web technologies [7] (2006)	A context-aware semantic e-learning approach to integrate content provision, learning process and learner personality in an integrated semantic e-learning framework.	It proposed to structure the semantics of contextual relations and concepts in various contexts, such as learning content description, learning model, knowledge object representation and learner personality
Herwig Rollett*, Mathias Lux, Markus Strohmaier, Gisela Dösinger, Klaus Tochtermann	The web 2.0 way of learning with technologies [8] (2007)	This paper explains the background of Web 2.0, investigates the implications for knowledge transfer in general, and then discusses its particular use in eLearning contexts with the help of short scenarios.	

(*continued*)

Table 1. (*continued*)

Author	Title	Key issues	Contribution
Carsten Ullrich, Kerstin Borau, Heng Luo, Xiaohong Tan, Liping Shen, Ruimin Shen	Why Web 2.0 is good for learning and for research: principles and prototypes [9] (2008)	Paper suggests that Web 2.0 is not only suited for learning but also in research process.	Web 2.0 provides constructivism and connectionism to make e-learning more attractive
Stefano Ferretti et al.	E-learning 2.0: you are We-LCoME! [10] (2008)	Paper proposes an e-learning 2.0 tool We-LCoME designed and developed to support users in editing educational resources and compounding multimedia contents through a collaborative work.	The We-LCoME system allows the cooperative creation and sharing of SMIL-based multimedia resources
Liana Razmerita*, Kathrin Kirchner, Kathrin Kirchner	Personal knowledge management: the role of Web 2.0 tools for managing knowledge at individual and organizational levels [11] (2009)	Web 2.0 enables PKM model that facilitates interaction, collaboration and knowledge exchanges on the web and in organizations.	
Denise Wood and Martin Friedel	Peer review of online learning and teaching: Harnessing collective intelligence to address emerging challenges [12] (2009)	This paper reports on a collaborative project led by the University of South Australia, which designed and developed a comprehensive, integrated peer review system that harnesses the power of the collaborative web.	The project use following open source technologies: • Apache HTTP Server 2.0.63 • PHP (PHP: Hypertext Processor) • MySQL 5.0.51a
Mohamed Amine Chatti, Matthias Jarke and Marcus Specht	The 3P learning model [13] (2010)	The paper proposes a framework for social software supported learning, stressing on 3Ps: Personalization, Participation and knowledge Pull in learning model.	It explains the concepts of intelligent tutoring systems (ITS), PLE, PKN etc.
Shiu-Li Huang and Jung-Hung Shiu	A user-centric adaptive learning system for E-Learning 2.0 [14] (2012)	The publication proposes a user-centric adaptive learning system (UALS) that uses sequential pattern mining to construct adaptive learning paths for the learners.	It uses collective intelligence and employs Item Response Theory (IRT) with collaborative voting approach to estimate learners' abilities for recommending adaptive material

5 Collective E-library Site Architecture

The proposed E-library site will provide intelligence to the learner by grilling the knowledge base in an intelligent inference engine. The knowledge is either content based or collaborative based. For the success of an application, it is necessary to use a right architecture. Therefore, we begin by describing the architecture for the proposed system. This will be followed by the services provided by the system and then content and collaborative filtering will be explained and finally, how the intelligence is represented and inferred for the learner.

The paper uses Event driven Service Oriented Architecture (SOA) and is composed of well defined synchronous services and asynchronous services referred in [3].

Synchronous service can be explained as services working on request/response basis. The service's response time should be less, reducing the client's (Learner) waiting time. Such services are user interacting example providing results from a search query for the learner or entering user profile. All the information these service needs to process is retrieved from a persistent source such as database.

Asynchronous service runs in the background and not time constraint. They receive a message, process it and then works for next. Load balancing is used in message server for controlling multiple instance of the same service queued up. Such services provided by the application will be:

(a) User Profile Analyzer, it includes background learning and analyzing the users proficiency, interest, skills through his profile and interaction activities in the background.
(b) Core Database creation and maintenance, it includes creating, structuring, aggregating and updating data/contents in the knowledge base by authoring experts.

5.1 Working of the Proposed Collective E-library Application

Based on the architecture, the Stepwise detailed system components operations:

Step1: The new user has to register on the application, through which the system collects learner personal details and store it in the user account database (Synchronous service). This profile details helps the system to predict learner's ability and choice.

Step 2: The learner profile is referred by the intelligent learning service when recommending learning material to the learner (Asynchronous service).

Step 3: The intelligent agent at the background maintains the authenticity and up-gradation of the material, refer Fig. 3.

Step 4: Depending on the learner feedback, her profile is regularly updated. This helps in the correctness and accurateness of the results (Asynchronous service), refer Fig. 4.

Step 5: When learner login the application, she enters her topic to be searched. Depending upon her interest/expertise in the topic and recommendation of the people of similar like hood and expertise, appropriate results are shown, refer Fig. 4

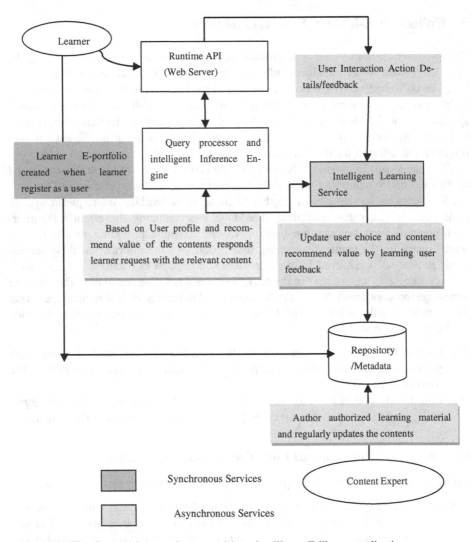

Fig. 3. Architecture for event driven intelligent E-library application

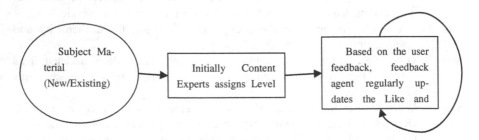

Fig. 4. Learning process for the quality and relevance of the content

Step 6: When learner completes reading the material, a feedback is taken from the
         learner. Based on the feedback, the material's readability and content level is
         updated, refer Fig. 4

Step 7: Learner is also given a platform to write reviews about the content. There
         are blog rooms also for each content, where user can query and discuss the
         relevant topics of the content with the co learners and the experts (Fig. 5).

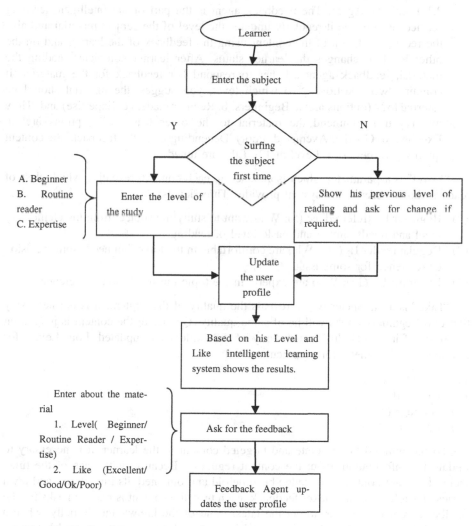

Fig. 5. Runtime query processing and feedback system of the application.

5.2   Components of the Proposed Application (Collective E-library Site)

(1) Runtime Interface: Runtime Interface enables learner to login, create her profile, Query searching and interact with system. It provide user with space to write reviews and suggestions about the content of the material studied. It also provides user with the blog space to interact with the other learners and experts.

(2) Intelligent Learning Service

2.1 Feedback Agent: The Feedback agent is the part of the Intelligent learning service, at one side it regularly updates the level of the subject material and also the recommendation of the content using the feedback of the learner and on the other hand, it changes the learner status. After learner completes reading the material, feedback agent ask her to respond her feedback for the material. It contains two questions: "To which level, you suggest the material should be referred to?" (options are: a. Beginners, b. Regular readers c. Expertise) and "How much you recommend the material to the other learners?" (options are: a. Excellent b. Good c. Average d. Poor). Depending upon the feedback, the content quality parameters: a. Level and b. Like are tuned.

"Level" is a parameter which is attached to each content, revealing which class of readers should refer it. the system provides Three levels:

(1) Beginner L_1 (referred as: (5): Who want to study the subject from the elementary level and needs simple and basic level of reading.

(2) Regular reader L_2 (10): Who are comfortable in the topic, but needs some revision or reference for some issue.

(3) Expertise L_3 (15): Who are experts in the topic but needs some reference.

"Like" is a parameter which reveals the quality of the content. It is not necessary that two Beginners books will be of same quality. Quality of the content is judged on the basis of its readability, authenticity of the content and updated. Four Levels for judging Like parameter for a particular level:

(1) Excellent (15)

(2) Good (10)

(3) Average (5)

(4) Poor (0)

To recommend the accurate and targeted content to the learner, it is necessary to update the information about the content regularly. Because, with pace of the time, technology and content knowledge become old and outdated, its importance and level varies. Therefore, system takes feedback, each time the content is read and asks for the feedback from the learner, to get the latest trend of the knowledge. Initially, when a material is introduced in the system and has no readers yet, its Level and Likes are initialized by the content experts of the system. But after that, each time the material is referred, its Level and Likes are tuned using collaborative voting approach [17]. This approach helps in reducing Expert Biasing. Tuning is required to balance between diversity and expertise. Many times diversity of learner can skew the results. So some fraction of expertise decision and some fraction of collaborative learner's feedback is considered [18].

Average Level of each material is calculated based on collaborative voting

$$Level\,(voting) = \sum\nolimits_{i=1}^{3} \left(\frac{n_i}{N}\right) * L_i, \qquad (1)$$

where level is the average Level parameter for the content,
n_i is the number of Learners that give feedback response to i^{th} Level
N_i is the total number of Learners who have give feedback for that content
L_i denotes (L_1, L_2, L_3)

Tuned Level of subject material

$$Level\,(tuned) = w_1 \times Level\,(initial) + (1 - w_1) \times Level\,(voting), \qquad (2)$$

where w_1 is the weight age factor
Level (initial) is the level assigned by content expert initially
Level (voting) is calculated as above
Similarly,

$$Like\,(voting) = \sum\nolimits_{i=1}^{4} \left(\frac{n_i}{N}\right) * R_i, \qquad (3)$$

where level is the average Level parameter for the content,

n_i is the number of Learners that give feedback response to i^{th} Like
N_i is the total number of Learners who have give feedback for that content
R_i denotes (R_1, R_2, R_3, R_4)

$$Like\,(tuned) = w_2 \times Like\,(initial) + (1 - w_2) \times Like\,(voting), \qquad (4)$$

where w_2 is the weight age factor
Like (initial) is the recommendation assigned by content expert initially
Like (voting) is calculated as above
Parameters on which Level (tuned) and Like (tuned) depend depicted in Charts 1, 2, and 3.

For charts Level (tuned) is taken into account, the same results holds for Like (tuned) also:
a. The weightage factors w1 and w2,
Greater is the value w_1 and w_2, more is the weightage given to Expert comments and chances of Expert Biasing is there.
Lesser is the value w_1 and w_2, more is the weightage given to the Collective Intelligence and chances of Skewness in value due to diversity in learners.
b. Number of Learners giving feedback (Honest and Helpful)
Therefore,

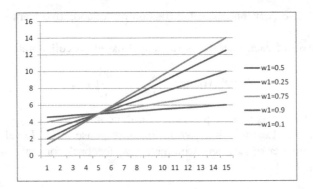

Chart 1. X Axis depicts Level (voting) Y Axis depicts Level (tuned) for given Level (initial) = 5 with w_1 = 0.5, 0.25, 0.75.0.9, 0.1

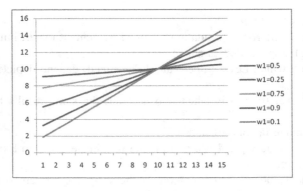

Chart 2. X Axis depicts Level (voting) Y Axis depicts Level (tuned) for given Level (initial) = 15 with w_1 = 0.5, 0.25, 0.75.0.9, 0.1

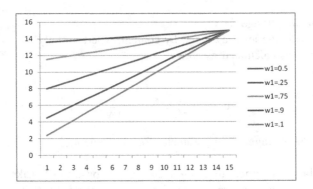

Chart 3. X Axis depicts Level (voting) Y Axis depicts Level (tuned) for given Level (initial) = 10 with w_1 = 0.5, 0.25, 0.75.0.9, 0.1

To handle the chances of Skewness and Biasing, feedback system takes following steps:

a. To handle Skewness, value of weightage factors is kept:

$w_1 > = 0.5$

$w_2 > = 0.5$

Skewness can be defined as the asymmetry in the predicted value due to uneven distribution of value.

The value 0.5 helps to control the value of tuned value to be effected much in case of diverse views.

As Biasing can be checked and rectified easily but to identify skewness is difficult because feedback is purely dependent on Learner's proficiency and personal experience.

b. To handle biasing, The values of Like (voting) and Level (voting) are considered and checked with Like (initial) and Level (initial) (The values which are specified by the experts).

If there is a variation between Level (initial) which is defined by subject expert and Level (voting), defined by

Variation = Mod [Level (initial) – Level (voting)/Interval] is equal or more than 50% then the resource material is referred to another expert of the respective field for providing Level (initial). This will help to rectify expert biasing and will produce genuine results.

2.2 Intelligent Inference Engine: Intelligent Inference Engine is an intelligent Query processor, which after evaluating the request of the Learner, reveals the necessary, targeted and user centric material to the learner based on appropriate value of Level and Like. The inference engine takes the information from feedback agent to find the appropriate response. The inference engine finds the material of the particular level entered by the learner and then according to the recommendation of the co learners, the results are shown in sorted form.

(3) Repository/Metadata: Repository is the backend database of the system. All the materials and related information is saved in it. It is majorly composed of User Profile database and Course Content database. When user creates her profile, all details are saved on to the user profile database. Learner's profile is regularly updated based on the Learner's feedback. On the other hand, all the courseware material with their necessary metadata is saved in the course content database. The courseware collection is completely under the supervision of course experts.

6 Case Study and Performance of the Algorithm for the Feedback Model

Following are the six crucial cases which are generated to judge the performance of the algorithm (Table 2):

Table 2. Case study for testing the system performance

S. no	Weightage factor w_1	Level (initial) by Expert	Level (voting) by feedback	Level (tuned) includes Level (initial) and Level (voting)	Action20
1	0.5	10	100 Learners gives 10 20 Learners gives 5 Level (voting) = (100*10) + (20*5)/120 = 9.1	Level (tuned) = 0.5*10 + 0.5*9.1 = 9.55	As Difference Between Level (initial) and Level (voting) is (10–9.1)/5 = 0.18 (or 18%) Therefore, It's the best case which no biasing and learner's diverse view
2	0.5	10	20 Learners gives 10 100 Learners gives 5 Level (voting) = (100*5) + (20*10)/120 = 5.8	Level (tuned) = 0.5*10 + 0.5*5.8 = 7.91	As Difference Between Level (initial) and Level (voting) is (10–5.8)/5 = 0.88 (or 88%) Therefore, Second Expert will consider for second opinion If Learner's view is genuine (expert biasing), other expert will alter level (initial) If Expert is correct, second expert will keep the old value
3	0.5	10	100 Learners gives 10 200 Learners gives 5 Level (voting) = (100*10) + (200*5)/300 = 6.67	Level (tuned) = 0.5*10 + 0.5*6.67 = 8.3	As Difference Between Level (initial) and Level (voting) is (10–6.67)/5 = .67 (or 67%) Therefore, Second Expert will consider for second opinion
4	0.9	10	10 Learners gives 10 200 Learners gives 5 Level (voting) = (100*10) + (200*5)/300 = 6.67	Level (tuned) = 0.9*10 + 0.1*6.67 = 9.667	As Difference Between Level (initial) and Level (voting) is (10–6.67)/5 = .67 (or 67%)

(continued)

Table 2. (continued)

S. no	Weightage factor w_1	Level (initial) by Expert	Level (voting) by feedback	Level (tuned) includes Level (initial) and Level (voting)	Action20
					Therefore, Second Expert will consider for second opinion But Temporarily tuned value is biased by expert value.
5	0.1	10	10 Learners gives 10 200 Learners gives 5 Level (voting) = (10*10) + (200*5)/ 210 = 5.2	Level (tuned) = 0.1*10 + 0.5*5.2 = 3.6	As Difference Between Level (initial) and Level (voting) is (10–5.2)/ 5 = 0.96 (or 96%) Therefore, Second Expert will consider for second opinion But Temporarily tuned value is biased by learner's choice
6	0.1	5	100 Learners gives 10 100 Learners gives 5 100 Learners gives 15 Level (voting) = (10*100) + (100*5) + (100*15)/ 300 = 10	Level (tuned) = (0.1*5) + (0.9*10) = 9.5	As Difference Between Level (initial) and Level (voting) is (10–5)/5 = 100% Therefore, Second Expert will consider for second opinion But Temporarily tuned value is biased by learner's choice and value is also skewed due to huge variation in feedback

7   Benefits and Challenges of the Social E-learning

As discussed previously, there will be lots of Benefits of using Collective Intelligence in E-learning such as [18]:

1) Learning will be social, open, emergent, driven by knowledge Pull.
(2) There will be less chance of malicious content, as distributed surveillance helps in better error detection."With enough eyes, all bugs are shallow", its well said saying, stating that more are the reviewers, more are the chances of revealing the content quality.
(3) As expert assessment can be biased, the content interpretation is subject to various perspectives, resulting in more balanced judgment.
(4) The application follows Additive Aggregation, that is taking feedback of all and then averaging the response.

But still there are lots of challenges in front of the success of the application such as [18]:

(1) Learner modeling is an extremely difficult task, due to the dynamics of the learner's knowledge and the diversity of the parameters that should be taken into consideration, such as the context of the learning environment, the nature of the learning activity, learning goals, preferences, motivation, cognitive capacities, disabilities, etc. (Aroyo et al., 2005).
(2) There should be a balance between diversity and Expertise. Many times diversity in learners can skew the results.
(3) It is also required that people involved in feedback should be intellectually competent to give a helpful and honest feedback. Therefore, application is creating user profile to judge her proficiency.
(4) User should have a desire to share knowledge or share experience and a sense of civic duty. They should be responsible, honest and capable for the knowledge, ratings and their views.
(5) The crowd's collective intelligence will only provide better results if individuals are open, diverse and derived collective intelligence can be properly grilled and utilized in decision making process.

References

1. Gruber, T.: Collective knowledge systems: where the social web meets the semantic web. J. Web Semant. Sci. Serv. Agents World Wide Web **6**(1), 4–13 (2008)
2. Ruttenbur, B.W., Spickler, G.C.: E-learning Book- the Engine of the Knowledge Economy, pp 16–20. Morgan Keegan and Company, New York (2000)
3. Alag, S.: Gathering data for intelligence. In: Collective Intelligence in Action, pp 30–46. Manning publications Co, Greenwich (2009)

4. Görgün, I., Türker, A., Ozan, Y., Heller, J.: Learner modeling to facilitate personalized E-learning experience. In: Kinshuk, D.G., Isaías, P.T. (eds.) CELDA 2005: Cognition and Exploratory Learning in Digital Age. International Association for Development of the Information Society (IADIS), pp. 231–237 (2005). http://wundt.unigraz.at/publicdocs/publications/file1170168718.pdf

5. Chen, C.-M., Lee, H.-M., Chen, Y.H.: Personalized e-learning system using item response theory. Comput. Educ. **44**, 237–255 (2005)

6. Teo, C.B., Gay, R.K.L.: Driven model to personalize E-learning. J. Educ. Resour. Comput. **6** (1), 3 (2006)

7. Huang, W., Webster, D., Wood, D., Ishaya, T.: An intelligent semantic e-learning framework using context-aware Semantic web technologies. Br. J. Educ. Technol. **37**(3), 351–373 (2006)

8. Rollett, H., Lux, M., Strohmaier, M., Dösinger, G., Tochtermann, K.: The Web 2.0 way of learning with technologies. Int. J. Learn. Technol. **3**(1), 87–107 (2007)

9. Ullrich, C., Borau, K., Luo, H., Tan, X., Shen, L., Shen, R.: Why web 2.0 is good for learning and for research: principles and prototypes. In: Proceeding WWW 2008 Proceedings of the 17th International Conference on World Wide Web, pp. 705–714 (2008)

10. Ferretti, S., Mirri, S., Muratori, L.A., Roccetti, M., Salomoni, P.: E-learning 2.0: you are WeLCoME. In: Proceedings of the 2008 International Cross-disciplinary Conference on Web

11. Razmerita, L., Kirchner, K., Sudzina, F.: Personal knowledge management: the role of web 2.0 tools for managing knowledge at individual and organizational levels. Online Inf. Rev. **33**(6), 1021–1039 (2009)

12. Wood, D., Friedel, M.: Peer review of online learning and teaching: harnessing collective intelligence to address emerging challenges

13. Chatti, M.A., Jarke, M., Specht, M.: The 3P learning model. Educ. Technol. Soc. **13**(4), 74–85 (2010)

14. Huang, S.-L., Shiu, J.-H.: Adaptive learning system for E-learning 2.0. Artif. Intell. Rev. **44** (3), 365–391 (2015)

15. Goldberg, D., Nichols, D., Oki, B.M., Terry, D.: Using collaborative filtering to weave an information tapestry. Commun. ACM **35**(12), 61–70 (1992)

16. Chen, C.M., Lee, H.M., Chen, Y.H.: Personalized e-learning system using item response theory. Comput. Educ. **44**(3), 237–255 (2005). Elsevier

17. Surowiecki, J.: The Wisdom of Crowds. Random House, Inc., New York (2005)

18. Bonabeau, E.: Decisions 2.0: the power of collective intelligence. MIT Sloan Manag. Rev. **50**(2), 45–52 (2009)

Behavioral Study of Defocus Cue Preserving Image Compression Algorithm for Depth Perception

Meera Thapar Khanna[1(✉)], Moumita Bhowmick[3], Santanu Chaudhury[1,2], Brejesh Lall[1], and Amrita Basu[3]

[1] Multimedia Lab, Department of Electrical Engineering,
Indian Institute of Technology Delhi, Delhi, India
meerathapars@gmail.com, schaudhury@gmail.com, brejesh.lall@gmail.com
[2] CSIR-Central Electronics Engineering Research Institute, Pilani, India
[3] Neuro Cognition Lab, School of Cognitive Science,
Jadavpur University, Kolkata, India
bhowmickmoumita1@gmail.com, amrita8@gmail.com

Abstract. Image and video processing is currently active research field during the past few years. Different coding schemes are available in the literature for image and video compression to improve compression ratio while maintaining picture quality. Many of the algorithms use ROI coding such as saliency based concept using different image features. But very few works related to depth cues preserving compression. In this paper, we present a behavioral study to state that the images compressed using defocus cue preserving compression yields better depth perception as compared to standard JPEG compression. We compare images compressed using different schemes against the original image. We collect data from different participants by showing original and compressed images to them. The responses are analyzed using analysis of variance. The analysis shows that the images compressed using defocus cue based compression provides the better perception of the raw image as compared to standard JPEG compressed image.

Keywords: Behavioral study · Defocus cue · Depth perception · Video coding · Depth cue preservation

1   Introduction

The depth perception is the visual ability to determine the distance of an object and see the world in 3D. The human eye has the ability to perceive depth from single still images by using various monocular and binocular depth cues such as defocus, texture, motion parallax, disparity etc. For instance, in our daily life, when we look around like in a stadium, the chairs with fine details and larger in size appears to be closer as compared to the chairs with coarser details and smaller in size. We can infer the depth information from the apparent variations

© Springer International Publishing AG 2017
A. Basu et al. (Eds.): IHCI 2016, LNCS 10127, pp. 264–275, 2017.
DOI: 10.1007/978-3-319-52503-7_21

in the local structures of textures in an image. As another instance, shading cue is used to perceive shapes of various objects in 3D. When we draw a circle on a page it looks like a circle only, but on adding shading it gets volume and circle turn into a sphere. After giving different levels of shading to different spheres, it provides information about the order of spheres nearer to farther from the observer. Like these examples, there are various other depth cues available in the 2D images or videos for depth perception. Most of the work in this field focus on depth estimation only. But very few use the concept of using these depth cues as saliency for compression purpose.

There are various coding standards available for compression of the digital multimedia content. Standard JPEG compression technique [1] uses same quantization parameters for the entire image or video frame, therefore, results in loss of information which is more relevant to the observer. A compression algorithm proposed in [2] uses an edge-based representation because the human visual system is highly sensitive to edges. In [3], authors proposed a binocular integration behaviors model for the quality measurement of stereoscopic 3D images. There are various other compression schemes that use saliency maps also for non-uniform bit allocation to the images or videos. Different approaches are available for saliency computation based on low-level or high-level features of images. There is a pixel domain method to compute saliency map by using some biologically inspired features extracted from the images [4]. Another method is to integrate both bottom-up and top-down attention models for optimal object detection [5]. In [6], authors proposed a method that works in the transform domain by capturing deviances while suppressing frequently occurring features in the images. Even though there are various saliency-based methods available for compression generating high visual quality, but it leads to the distortion of different cues because of the quantization process, which does not consider the significance of these cues. These cues are important from the depth perception point of view. However, the compression engines do not address this aspect of the problem to incorporate these cues as saliency during the compression process. Therefore, there is a need to address the problem of preserving various depth cues in images or videos for depth perception.

Defocus is one of the important monocular depth cues. It indicates some part of the image is sharply focused than another part. In [7], authors proposed a different way that shows the defocus blur might affect depth perception. Because it is a cue to egocentric distance and could contribute to quantitative depth perception. In [8], authors proposed a method for image compression to retain depth cue in the images. In this work, ROI (region of interest) based compression scheme is used for non-uniform bit allocation to the images while compression. Initially, defocus saliency maps are computed using the frequency information in the image. Then these saliency maps were used for non-uniform bit allocation to the images.

Since humans are the end users in most of the multimedia applications. Therefore, subjective evaluation of image quality assessment is the most accurate and reliable method. In this paper, we present a behavioral study by rating images for the perception of images compressed using standard JPEG [1] and defocus cue preserving [8] compression algorithm. Here we present a study to validate the importance of defocus cue preserving compression scheme. We compare the images after compressing with defocus cue preserving scheme with the images after compressing with the standard JPEG scheme against the original images. This study is conducted by a group of people. They are asked to give their ranking to the images on the basis of similarity between original and compressed image. Analysis results show that there is a significant difference between original, standard JPEG and defocus cue based compressed images. Defocus cue based compressed images are better in retaining depth and provides a close perception to the raw image than standard compression method.

The remainder of the paper is organized as follows. Section 2 describes the different types of images used for the study and the step by step algorithm for the behavioral study. Section 3 describes the theoretical model and hypothesis to be tested for analysis. Section 4 presents result using analysis of variance and conclusions are drawn in Sect. 5.

2   Proposed Behavioral Study

An experiment is conducted on 20 subjects each labeled as 'a', 'b', ..., 't' to see the effectiveness of defocus cue preserving image compression scheme. All the subjects are within the age group of 20–35. The experiment is designed in psychological software tool e-prime 2.0 [10]. We used the raw image data set RAISE [9], which consists of high-resolution raw and uncompressed images. All the images belong to different categories such as indoor and outdoor scenes, nature and landscape scenes, people, objects and building photos. We compress these images using standard JPEG compression and defocus cue preserving saliency based algorithm at different degrees of compression such as at quality parameter from 3 to 10. These parameters are labeled as 'A', 'B', 'C', 'D', 'E', 'F', 'G', 'H'. The experiment is designed for all the quality parameters. Each experiment consists of 45 original images along with compressed images. The subjects are shown both original and compressed images using standard JPEG and defocus cue preserving saliency based compression scheme. The images are labeled as 'Raw', 'Standard_JPEG' and 'Saliency_JPEG' respectively. The subjects are totally incognizant of the experiments. They are required to assign ranks to images based on the similarity between original and compressed images for every level of the quality parameter. Each trial lasted for about 10–15 min. A three-way ANOVA analysis is then conducted on these 20 participant's ratings of similarity of images to examine the variation in perceptual responses arising due to different categories of images, different image qualities, and individual perceptions. A brief discussion for different compression schemes are as follows:

2.1   Standard JPEG Compression

JPEG standard is a commonly used method of compression for digital images [11]. We can adjust the degree of compression by selecting different quality parameters. So there is a tradeoff between storage size and image quality. JPEG compression is based on the discrete cosine transform (DCT). DCT operation converts each image from the spatial domain into the transform domain. Transform domain provides a convenient way to represent images. In transform domain, high-frequency information is discarded such as sharp transitions in intensity, and color hue. It works on 8×8 blocks. JPEG uses a fixed quantization table for the whole image. Information is lost during the quantization process. The high-frequency coefficients that contribute less to the overall picture than other coefficients are quantized during this process. These quantized coefficients are packed to generate the output bit-stream. JPEG is a form of lossy image compression, therefore, some of the information is lost during compression process and cannot be restored. This affects the quality of an image.

2.2 Defocus Cue Based Compression

We use defocus depth cue as a saliency measure for preserving defocus cue during the compression scheme [8]. We need to distinguish between the region of focus and rest of the image. Frequency information is a good way to measure the blurriness present in an image. In the image, the high-frequency component provides the edge information. However, there is some noise present in the high-frequency component of the image. Therefore, to reduce this noise DoG (Difference of Gaussian) kernels are used. In an image, the maxima of highest frequency values correspond to the distance of the edge most robustly. To reduce noise box filter is applied, which accept maxima over a certain fraction.

$$F(x,y) = min(f|S'(f)\ is\ a\ local\ maxima > \gamma) \tag{1}$$

where $F(x,y)$ is the box filtered output for a given pixel, that is a slice through the DoG kernels. We experimentally obtain the value of γ and it comes out to be 0.09. Frequency information provides the saliency map for edge pixels only but we require saliency information for the entire image. Therefore, this focal information is transferred from edge pixels to non-edge pixels to form the final saliency map.

Loss occurs in JPEG compression during the quantization step. In JPEG compression a quality factor is used to decide the level of compression. Low-quality factor results in high compression and loss of some important information in an image. This quality value is usually fixed for the complete image. Here we modify the quality value for each block based on the factor that how salient is that block from the depth point of view. Therefore, there will be one universal quality value which will determine the quality of the compression and an additional value for each 8×8 block. The Quality value is calculated using the saliency map. We consider the multiplicative factor to be the square root of the average amount of saliency present in that particular block. The formulation for quantization value for a particular block is given as:

$$QP_{block(i)} = Q * \sqrt{\frac{Mean_{sal}(Block(i))}{Mean_{sal}(Image)}} \tag{2}$$

where Q denotes the quality factor, $QPBlock(i)$ denotes the quality factor for block i, *Image* is the entire image.

This process takes less time because quality factor used for quantization is computed on the fly separately for each block. We need to transmit this quality factor along with the image, as it is not a part of the JPEG coding scheme.

2.3 Algorithm of the Behavioral Study for Rating the Images

The step by step process for evaluation is as follows:

1. There is a welcome screen to specify the instructions for the experiment.
2. Initially, a sample (Original/Raw) image is shown to the viewer.
3. Then a filler screen appears having + sign in it.
4. Filler screen is followed by a target image.
5. Target images can occur in four different combinations:
   (a) First, sample image and standard JPEG compressed image as target,
   (b) Second, sample image and defocus cue preserving saliency based compressed image as target,
   (c) Third, sample image and same sample image as target and,
   (d) Fourth, sample image and different sample image as the target image.
6. Then there is a screen to take the response from each subject/viewer. The subjects have to assign the similarity on the scale of 1 to 7.
   (a) If the sample and target images appear completely similar, then press 1,
   (b) If images appear totally different to the subject then press 7,
   (c) Otherwise rank between 2 to 6 depends on the difference between the sample and target images.

The fourth case in step 5 acts as a vigilance check when different sample image comes as the target against sample image to have the high accuracy of results.

3 Behavioral Study Analysis

A Three-way ANOVA is conducted to compare the subject responses between different compression schemes with quality parameter, image type and subject as the three independent factor and individual responses as the dependent factor. The theoretical model for our analysis is as follows:

3.1 Theoretical Model

Assumptions. The following are the assumptions:

1. The observations are random samples from the normal distribution.
2. The population has the same variance σ^2.
3. Observations are independent.

Linear Model. The following linear model is defined for three way analysis of variance:

$$Y_{ijkl} = \mu + \alpha_i + \beta_j + \gamma_k + (\alpha\beta)_{ij} + (\beta\gamma)_{jk} + (\gamma\alpha)_{ki} + (\alpha\beta\gamma)_{ijk} + \in_{ijkl},$$
$$i = 1, ..., I, j = 1, ..., J, k = 1, ..., K, l = 1, ..., L \tag{3}$$

where Y denotes the response variable, i denotes the type of image, categorized according to different compression schemes, j denotes the image quality, categorized according to different quality schemes undertaken by the subject, k denotes the subject, l denotes series of values within a cell; μ represents mean of all image percepts, α_i represents main effects of the rows i.e., the type of image, β_j represents main effects of the columns i.e., the quality of image, γ_k represents main effects of the subjects, $(\alpha\beta)_{ij}$ represents the interaction effect of image type and image quality, $(\beta\gamma)_{jk}$ represents the interaction effect of image quality and subject, $(\gamma\alpha)_{ki}$ represents the interaction effect of subject and image type, $(\alpha\beta\gamma)_{ijk}$ represents the interaction effect of image type, image quality, and subject, $\in_{ijkl}$ denotes the error term where $\in_{ijkl} \sim^{iid} N(0, \sigma^2)$.

$Y_{ijkl} \sim^{iid} N(\mu_{ijk}, \sigma^2)$, where μ_{ijk} is the theoretical mean of all observation in the 3-D space (i, j, k)

Remarks. Although the graphical analysis of the observations such as histogram and density plot reveals the data to be positively skewed with mode 1, median 2 and mean 2.5 but given that ANOVA is robust to the assumptions of normality and observed near symmetrical nature of the residuals, we went ahead with the analysis. The results of the graphical analysis and the summary of the descriptive statistic can be found in the appendix section.

3.2 Hypothesis to Be Tested

We define the following hypothesis to test:

H_{00}: There is no difference in the mean responses of images across types, i.e.

$$\mu_{1jk} = \mu_{2jk} = \mu_{3jk} = ...$$

H_{01}: There is differences in the mean response of images across types, i.e.

$$\mu_{1jk} \neq \mu_{2jk} \neq \mu_{3jk} \neq ....$$

H_{10}: There is no difference in the mean responses of images across image qualities, i.e.

$$\mu_{i1k} = \mu_{i2k} = \mu_{i3k} = ...$$

H_{11}: There is differences in the mean response of images across image qualities, i.e.

$$\mu_{i1k} \neq \mu_{i2k} \neq \mu_{i3k} \neq ....$$

H_{20}: There is no difference in the mean responses of images across subjects, i.e.

$$\mu_{ij1} = \mu_{ij2} = \mu_{ij3} = \ldots.$$

H_{21}: There is differences in the mean response of images across subjects, i.e.

$$\mu_{ij1} \neq \mu_{ij2} \neq \mu_{ij3} \neq \ldots.$$

H_{30}: There is no interaction between the image types and the image qualities
H_{31}: There is interaction between the image types and the image qualities
H_{40}: There is no interaction between the subjects and the image types
H_{41}: There is interaction between the subjects and the image types
H_{50}: There is no interaction between the image types, image qualities and the subjects
H_{51}: There is interaction between the image types, image qualities and the subjects

To test the hypotheses, we calculate F-statistic and compare it with the one-tailed critical F value from F distribution. If the null hypothesis is true, the observed F-value will be less than the critical F-value. On the other hand, if observed F is greater than critical F value for a given level of significance, we reject the null hypothesis and accept the alternative hypothesis.

4   Results and Discussion

Table 1 shows the ANOVA table obtained using 'R' statistical software. There is a significant main effect for 'Image type' (F = 83.1383, p-value <2.2e−16). In general, the perceptual responses of raw images (M = 1.84, SD = 1.82) are close to 1 compared to the other two types of images. This implies, the participants could discriminate the original images from the other two compressed images. Note that the perceptual responses of Saliency_JPEG images (M = 2.78, SD = 1.55) are closer to 1 as compared to the Standard_JPEG images (M = 3.11, SD = 1.74). This implies, the images which are compressed using defocus cue preserving compression algorithm provides better perceptual estimates of the original images. There is significant main effect for 'subject' (F = 3.2536, p-value = 2.063e−06). Perceptual variation in mean responses across the individuals is observed with participant h's overall response (M = 2.95, SD = 1.91) to be the highest and participant l's overall response (M = 2.20, SD = 1.21) being the lowest. There is no significant main effect for 'image quality' (F = 1.4125, p-value = 0.1951). There is a significant impact of image type × image quality on the perceptual responses (F = 12.4667, p-value <2.2e−16). This implies that the mean responses are jointly affected by the image type and the image quality.

It is clear from the Fig. 1 that the images in third and fifth column are better than second and fourth column images and more close to the original images. For instance, defocus cue is present in the flowers, branches or in the background of the first image. Standard JPEG compression scheme uses the same

Table 1. Summary table for three way ANOVA

	Sum of squares	Degree of freedom	F-obs	Pr > F	Critical F value (0.05) [12]
Intercept	98.5	1	51.7240	6.714e−13	
Image type	316.5	2	83.1383	<2.2e−16	3.00
Image quality	18.8	7	1.4125	0.1951	2.02
Subject	117.7	19	3.2536	2.063e−06	1.62
Image type: image quality	332.2	14	12.4667	<2.2e−16	1.72
Image type: subject	820.9	38	11.3488	<2.2e−16	1.46
Image quality: subject	1035.7	133	4.0912	<2.2e−16	1.21
Image type: image quality: subject	2261.9	266	4.4673	<2.2e−16	1.21
Error	26496	13920			
Total	31498.2	14400			

Fig. 1. (a) Original image, compression results using (b) Standard JPEG scheme (c) Defocus cue based scheme at quality parameter 3, compression results using (d) Standard JPEG scheme (e) Defocus cue based scheme at quality parameter 5.

quantization values for the entire image whereas the defocus cue based scheme assign quantization based on the defocus cue present in different regions of the image. Therefore preserve the defocs cue and provides better depth perception.

Figure 2 shows that the mean responses of the Raw image remains within the range of 1 and 2 across the image qualities. Note that, for Standard_JPEG and Saliency_JPEG images, the mean responses drop as the image quality moves from A to H (low to high quality). However, the mean response of Saliency_JPEG is lesser than the mean response of Standard_JPEG images across the image qualities.

Figure 3 shows that for all the image qualities (A to H i.e. 3 to 10) the mean responses increase from image category saliency_JPEG to standard_JPEG.

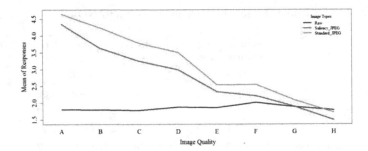

Fig. 2. Interaction plot of responses for the variables image quality and image type split by image quality.

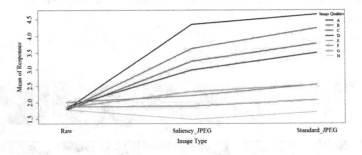

Fig. 3. Interaction plot of responses for the variables image quality and image type split by image type.

This implies that saliency_JPEG to be able to provide better percept of the raw image as compared to the standard_JPEG.

There is a significant impact of image type × subject on the perceptual responses (F = 11.3488, p-value <2.2e−16). This implies that there is a joint effect of individual perceptual variability and differing image types on the participant ratings.

Figure 4 shows that the behavioral responses increased from Raw image type to standard image type across all the individuals. Note that the mean behavioral responses of most of the individuals increase from image category saliency_JPEG to standard_JPEG. This implies Saliency_JPEG to be able to provide better percept of the original image as compared to Standard_JPEG images.

In Fig. 5, the variation in the level of mean responses across the subjects remains fairly similar for all the image categories. However, the mean response level for the saliency_JPEG image is lower than the standard_JPEG image type across all the individuals. There is a presence of individual variation in the perception of images.

There is a significant impact of image quality × subject on the perceptual responses (F = 4.0912, p-value <2.2e−16). This implies that there is a joint effect of individual perceptual variability and differing image quality on the perception of the images.

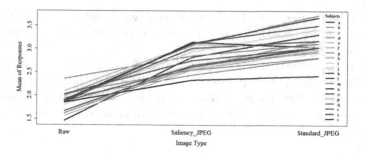

Fig. 4. Interaction plot of responses for the variables subject and image type split by image type.

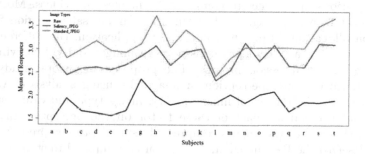

Fig. 5. Interaction plot of responses for the variables subjects and image type split by subjects.

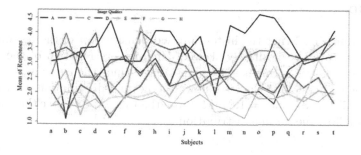

Fig. 6. Interaction plot of responses for the variables image quality and subjects by subject.

It can be observed from Fig. 6 that mean scores of the image quality A is higher as compared to the image quality H across the subjects. The mean score of the subjects' approaches 1 as the image quality parameter shifts from A to H (low to high quality). The degree of individual variability is also observed to be high for image quality A as compared to image quality H.

Thus, as the quality parameter increases from A to H, the subjects will have better clarity in understanding the images and hence the lesser perceptual variation in the mean scores. Lastly, there is a significant joint impact of image type, image quality and individual perception on the mean responses (F = 4.4673, p-value <2.2e−16).

Hence, we conclude that image perception is dependent not only on the image type but also on the quality parameter and it is susceptible to individual perceptual variability. Results show that the defocus cue preserving compression schemes yields better depth perception as compared to standard JPEG compression scheme.

5 Conclusion

There are various approaches to compress 2D images and videos. Most of the schemes use low-level or high-level features for finding salient region for non-uniform bit allocation during the compression application. But different depth cues are also important for depth perception in images and videos, which may lose during the compression process. The proposed behavioral study shows the importance of depth perception in images by using analysis of variance. The study shows the impact of image quality, image type, individual subject's response and the joint impact of these factor on the perception of images. This study shows that the images compressed using depth cue preserving algorithm is closer to the Raw image in perception as compared to standard JPEG compressed image.

Appendix

It is clear from Fig. 7 that the histogram and density plot of the response data reveals the data to be positively skewed. The density function of the residual is plotted to test the normality of the residuals. Since the curve is nearly symmetrical, we can conclude that it follows the normal distribution with mean = 2.885597e−17 and variance = 1.84014. From the Fig. 8, we can see that the residuals nearly follow the normal distribution.

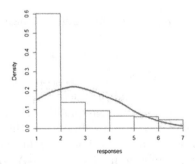

Fig. 7. Histogram and density plot of the responses.

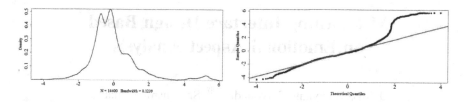

Fig. 8. (a) Density plot of the residuals of the model (b) Normal Q-Q plot and normal Q-Q line of the residuals.

References

1. Marcus, M.: JPEG image compression, https://math.dartmouth.edu/m56s14/proj/Marcus_proj.pdf
2. Desai, U.Y., Mizuki, M.M., Masaki, I., Horn, B.K.P.: Edge and mean based image compression (1996)
3. Lin, Y.H., Wu, J.L.: Quality assessment of stereoscopic 3D image compression by binocular integration behaviors. IEEE Trans. Image Process. **23**, 1527–1542 (2014)
4. Itti, L., Koch, C.: Feature combination strategies for saliency-based visual attention system. J. Electron. Imaging **10**, 161–169 (2003)
5. Navalpakkam, V., Itti, L.: An integrated model of top-down and bottom-up attention for optimal object detection. In: IEEE Conference on CVPR 2006, pp. 2049–2056, June 2006
6. Hou, X., Zhang, L.: Saliency detection: a spectral residual approach. In: IEEE Conference on CVPR 2007, pp. 1–8, June 2007
7. Vishwanath, D.: The utility of defocus blur in binocular depth perception. i-Perception 3, 541–546 (2012)
8. Khanna, M.T., Rai, K., Chaudhury, S., Lall, B.: Perceptual depth preserving saliency based image compression. PerMIn **2015**, 218–223 (2015)
9. Dang-Nguyen, D.-T., Pasquini, C., Conotter, V., Boato, G.: RAISE a raw images dataset for digital image forensics. In: ACM Multimedia Systems, Portland, Oregon, 18–20 March (2015)
10. https://www.pstnet.com/eprime.cfm
11. Wallace, G.: The JPEG still picture compression standard. Commun. ACM **38**, 30–44 (1991)
12. http://www.stat.purdue.edu/jtroisi/STAT350Spring2015/tables/FTable.pdf

M-Learning Interface Design Based on Emotional Aspect Analysis

Djoko Budiyanto Setyohadi[1(✉)], Sri Kusrohmaniah[2],
Efrans Christian[1], Luciana Triani Dewi[3], and Bening Parwita Sukci[3]

[1] Magister Teknik Informatika, Universitas Atma Jaya, Yogyakarta, Indonesia
djoko@mail.uajy.ac.id, efranschristian2@gmail.com
[2] Psikologi Dept., Universitas Gadjah Mada, Yogyakarta, Indonesia
koes_psi@ugm.ac.id
[3] Teknik Industri, Universitas Atma Jaya, Yogyakarta, Indonesia
triani.dewi@mail.uajy.ac.id, parwitasukci@yahoo.com

Abstract. Positive emotional conditions are influenced or generated by internal or external stimuli. Design elements, such as colours and shapes, are potential stimuli that can influence human emotion. The study is devided into two experimental stages. The first stage employs design elements, i.e. colours and shapes. The colours used are red, green, blue and yellow with round and angular shapes. While the second stage is testing the mobile learning prototype that applies the design elements as the results of the experiment in the first stage. The testing employs 41 SAM Questionnaire by measuring 41 respondents' perception towards design element stimuli and m-learning prototype and, later, the results are analyzed with the use of One-Way Anova. The results of the two experiments show that red, green and blue generate positive energy while yellow generates negative emotion. On the other hand, round shape can result in positive emotion and the angular shape is neutral.

Keywords: Positive emotion · Design elements · Mobile learning

1 Introduction

Emotion is a psychological and physiological condition as a reaction to stimulus received by human senses. Emotion can also be considered as a result of thinking illustration that takes place when there is an internal or external emotional stimulus [1]. Emotions are frequently considered as identical or closely related to feelings [2]. In general, emotions can be categorized into: positive and negative. The state and the role of one's positive emotion can influence the action or activity and will result in better learning [3]. When one is studying, the positive emotional state prior to the learning will help the person to understand better [4].

E-learning system that has been developed to enhance learning process can operate on desktop or mobile computer interface to enable users to interact with the system which employes design elements such as colours, shapes, sizes, etc. These factors can strongly generate a certain secondary emotion or aesthetical response when a person interacts with the interface system [5]. Among design elements stated above, colours and shapes have the most potential emotional influence on a person. Colours are frequently associated with positive and negative emotions [6, 7] depending on the sensation generated. On the other hand, other factors such as sexes, ages, and cultures can influence the perceptions and responses towards certain colours [8–10]. Colours can also influence attention, behaviors and achievements in a student's learning process [11]. Just as design elements like colour and shapes influence human emotion, shapes and characteristics such as round, angular, simplicity, and complexity can also influence human emotional response in visual arts and psychology [12].

Currently, there have been previous researches focusing on emotion and computer. However, mobile technology as a platform of m-learning is advantageous and it is still an infant that can develope further [13, 14]. Therefore, it is still necessary to research on how to develop m-learning, especialy in order to reduce the limitation of mobile technology. This limitation of the problem in this paper is to explore the use of mobile device particularly on human computer interaction. Shortly, this research studies colour and shape in design elements needed in m-learning by taking emotional aspects into consideration in designing interface mobile learning application.

2 Previous Work

This research refers to a previous research carried out by Abegaz [15], which tested the influence of the colour and shape design elements towards emotion in designing an interface search engine to improve comfort for adults when using the search engine. In his research, Abegaz used red and yellow and round and angular characteristics and also the combination of the two. Abegaz found out that colour and shapes able to influences emotions of the user.

Another research was carried out by Plass and friends to students in German University [16]. The results show that a combination of design elements of warm colours and round characteristics on multimedia learning interface can generate positive emotions, reduce difficulties and increase the motivation in learning. Certain colours, especially blue and yellow in the background can influence the process of learning through a computer. The results show that blue background generates a more positive influence than the yellow one [17].

The researches above employed different measuring instruments in accordance with the purposes of the study. The instruments are to measure emotional responses towards design elements such as colours and shapes. Furthermore, Self-Assessment Manikin (SAM) Questionnaire in measuring respondents' emotion perception towards the stimuli given via computer [18–20] is also used.

3 Research Methodology

Due to the aim of the research some definitions are used, i.e. colour, shape and SAM Questionnaire. Colour is a certain spectrum resulted from a perfect light. A colour is determined by the length of light. There are three main colours (red, green and blue), known as additive primary colours. The mixing of additive colours results in primary subtractive colours, such as magenta, cyan, and yellow [21]. Shape is a two dimension basic geometry, such as a dot, line, curve, field (i.e. square or circle). It can also be something explained by three dimension solid substance, such as a cube or a ball, or round and angular with simplicity or complexity [22]. SAM Questionnaire is a non-verbal picture oriented measuring technique that directly measures pleasure, arousal, and dominance related to a human affective reaction towards some stimuli [23].

Research design stage was caried out by studying the existing literature and previous study on colours and shapes [15]. It was held to get a reference on colours and shapes to use as models that would be tested and measured to design the interface mobile learning. Our research design is divided in two stages which are carried out sequentialy. To give clear explanation of both stages, the results and discussion will be described separately. This research carried out two experiments, the first is to measure and find out the emotional responses towards colour and shape design elements. The second is to measure and find out the emotional responses towards colours and shapes that have been shown as potentials in experiment one to generate states of positive emotion when applied on interface mobile learning. Each of the two experiments has four stages, research design, model design, data collecting and data analysis.

4   Methodology in Experiment One

The model design consists of two important parts employed as variables of the experiment i.e. shape and colour. Model designing is to create a colour and shape model used as a stimulus for the respondents. There are three models: colours only, shapes only, mix between colours and shapes. There are four chosen colours to test: red, green, blue and yellow. These four colours are combination of primary addictive and subtractive colours and considered as generating certain emotions [24]. The colour model also splits colours into some shades since colour dimensions such as hue, saturation and brightness potentially influence human emotions [25]. Shape Model is made by applying two characteristics of shapes, i.e. round and angular, and shapes that combine both characteristics.

Using random sampling, data were collected from 41 s semester Informatics Department students from batch 2015 of University of Palangkaraya Indonesia. There were 21 male and 20 female respondents ranging from 18–21 years old. The experiment was carried out indoors with sufficient light and air conditioner in the morning between 09.00–10.30 Indonesian Western Time. The purpose of this condition is to eliminate external factors that can influence the respondents' attention and emotional state while the data were taken.

Quasi was given to the 41 respondents to find their emotional tendency by using PANAS questionnaire as an instrument. The questionnaire that covers positive and negative effects is to determine emotional experience that each individual has had [26]. The quasi test shows that 87.8% or 36 respondents have positive emotional tendency while 12.2% or five respondents experience negative tendency. The experiment data is collected by using android smartphones to present colour and shape models and applying SAM Questionnaire to measure their emotional responses towards the stimuli. SAM Questionnaire was applied especially to measure pleasure dimension (see Fig. 1).

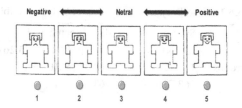

Fig. 1. Instrument SAM Questionnaire to measure pleasure dimension [23]

SAM Questionnaire used expression illustration to represent the states of emotion from negative, neutral and positive in showing pleasure dimension. The process of data collecting was carried out in three stages as shown in the following Fig. 2. The first and the second stages were in the form of application while the second, rating the emotion, was paper-based by using SAM Questionnaire. The first stage shows the models to the respondents within 1000 ms. This duration is considered sufficient to generate respondents's emotions either when they were aware or not [27]. In the second stage, after looking at the models in stage one, respondents were given SAM Questionnaire to choose one of the five available pictures that represents their emotion when they saw the model. In the third stage, the respondents were to look at the running application, in this stage the application did not present models, but a white blank screen for 2000 ms. The blank white screen is to neutralize the given stimuli to the eyes after looking at models in stage one. After stage three, the experiment proceeded to stage one to evaluate the following models. These three stages were repeated until all the models were displayed and given score by the respondents.

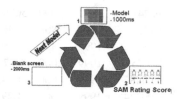

Fig. 2. Stages on data collecting process [15]

4.1 Data Analysis and Results of Experiment One

The data analysis employs One-Way Anova (OWA) to test the mean score comparisons that each model has achieved. The models are analised by using OWA to see which colours and shapes that give positive, negative or neutral emotions to respondents. Prior to data analysis, reliability test is carried out to the questionnaire data. It is necessary to make sure that all questionnaire items are consistent and reliable. The reliability test of 19 item model questions given to 41 respondents, is done by using Cronbach Alpha test. The reliability shows the score of Cronbach Alpha (0.812). By using *r-table* vales for *df* = 39 (0.3081) (*significant level* 0.05), the questionnaire is

Table 1. One-Way anova analysis of colour variable

	Sum of squares	df	Mean Sq.	F	Sig.
Between groups	32.604	3	10.868	13.261	.00
Within groups	131.122	160	.820		
Total	163.726	163			

reliable since the value of Cronbach Alpha (0.8012) > *r-tabel* (0.3081). Furthermore, the result OWA of the collected data can be seen in Table 1.

Then, comparative study was carried out to see the significant gap between the mean values by employing LSD (least significant difference) post-hoc test. The results showed that three colour models: red, green and blue have significant average gap towards yellow since the comparation of the value is $sig(0,000) < 0,05$, so it can be concluded that red, green and blue models have different mean emotion rating value towards yellow (let see Tables 2 and 3). On the other hand, red, green and blue models have no significant emotion rating mean scores since the three have values of *sig* $(0.465) > 0.05$, $sig(0.331) > 0.05$ dan $sig(0.808) > 0.05$.

Table 2. Post hoc test LSD of colour variable (with 95% Conf Int)

Colour 1	Colour 2	Mean diff.	Std Err.	Sig	Low bnd.	Upp bnd.
Red	Green	.1463	.1999	.465	−.2485	.5412
	Blue	−.0487	.1999	.808	−.4436	.3461
	Yellow	1.0487*	.1999	.000	.6539	1.4436
Green	Red	−.1463	.1999	.465	−.5412	.2485
	Blue	−.1951	.1999	.331	−.5900	.1997
	Yellow	.9024*	.1999	.000	.5076	1.2973
Blue	Red	.0487	.1999	.808	−.3461	.4436
	Green	.1951	.1999	.331	−.1997	.5900
	Yellow	1.0975*	.1999	.000	.7027	1.4924
Yellow	Red	−1.0487*	.1999	.000	−1.4436	−.6539
	Green	−.90244*	.1999	.000	−1.2973	−.5076
	Blue	−1.0975*	.1999	.000	−1.4924	−.7027

Table 3. Descriptive analysis of colour variable

Variable	N	Min.	Max	Mean	Std Dev.	Emotion
Red	41	3	5	4.0244	.7241	Positif
Green	41	2	5	3.8780	.9272	Neutral/Positif
Blue	41	3	5	4.0732	.8772	Positif
Yellow	41	1	5	2.9756	1.0603	Negatif/Neutral

The results show that yellow is able to stimulate negative emotion. Possibly it reflects the sad-perception of Dayak culture in Central Kalimantan since they use yellow flag in the Tiwah, a ritual held for the deceased.

As decribed before, we made use of two (2) variables in experiment one. When we used shaped variable, the result shows that there is no emotion affected by shape variable. It is seen from OWA analysis when the value of $F_{measure}$ as which is as much as 1,057 will, then, be compared with the scores of F_{table} (2;120; 0,05) which values 3.07. It is found that $F_{measure} < F_{table}$ and the comparative value sig (0.351) > 0.05. Thus we will not apply LSD post-hoc test. It reflects that the three shape models have the same average emotional rating. Shortly, variable shape has no effects on the emotion as seen on Table 4.

Table 4. Descriptive analysis of shape variable

Variable	N	Min.	Max	Mean	Std Dev.	Emotion
Angular	41	2.00	5.00	3.3659		Neutral
Round	41	2.00	5.00	3.6098		Neutral
Mix	41	1.00	5.00	3.3415	1.03947	Neutral

The next experiment explores on how both of shape and colour variables affect emotions. After the data are collected and processed, One-Way Anova Analysis of the Combination of Shape and Colour is performed. Table 5 shows results of this experiment. $F_{measure}$ (4,196) $< F_{table}$ (11;480;0,05), and the comparative value sig (0,000) < 0,05). Therefore we need to apply post-hoc test LSD.

Table 5. One-way anova analysis of combination of shape and colour

	Sum of Sq.	Df	Mean Sq.	F	Sig.
Between groups	44.998	11	4.91	4.196	.00
Within groups	467.902	480	.975		
Total	512.900	491			

Due to the limitation of space, we can only show the OWA analysis of the combination of colour and shape variables (let see Table 5), and the final result, a descriptive analysis of both variable, is presented in Table 6, while Post Hoc Test LSD is not. The summary of the table can be also used to confirm the conclusion that yellow has strong effect on the emotion.

Based on One-Way Anova Analysis of Combination of Shape and Colour and the LSD post-hoc test, the results of the mix models can be taken. The LSD post hoc test showed that there is significant difference between mixed models all those of the colours (red, green, blue) with the shapes (angular, round, mixed), yellow, and shape (angular, round and mixed. The results of the mixed model show that blue and round edge can potentially generate positive emotion in the respondents since it gets the highest mean (3,8780) although the standart deviation is 0,92723. While yellow combined with all kinds of edge tend generate negative emotions. The mixed models effects and the result analysis of this experiment is displayed in Table 6 as below.

Table 6. Descriptive analysis of combination of shape and colour

Variable	N	Min.	Max.	Mean	Std Dev.	Emotion
RedAngular	41	1.00	5.00	3.5854	1.0948	Neutral
RedRound	41	2.00	5.00	3.6098	.8330	Neutral
RedMix	41	1.00	5.00	3.4634	1.0270	Neutral
GreenAngular	41	2.00	5.00	3.7317	.9226	Neutral/Positive
GreenRound	41	2.00	5.00	3.6585	.8249	Neutral
GreenMix	41	2.00	5.00	3.5854	.9993	Neutral
BlueAngular	41	1.00	5.00	3.7561	.9945	Neutral/Positive
BlueRound	41	1.00	5.00	3.8780	.9272	Neutral/Positive
BlueMix	41	1.00	5.00	3.5854	1.0241	Neutral
YellowAngular	41	1.00	5.00	2.9512	1.0235	Negative/Neutral
YellowRound	41	1.00	5.00	3.0244	.9614	Neutral
YellowMix	41	1.00	5.00	3.0000	1.1619	Neutral

5 Methodology in Experiment Two

According to the purposes of the experiment two, three applications of m-learning prototype are designed and used to collect data. The result of experiment one is applied to select the design element of the prototype. Since this research is limited on mlearning, the interface mobile learning application prototype is limited on the standard learning activities, i.e. page on learning material selection, page on learning material presentation, and page on learning evaluation process via test or quizes. However in order to enrich the prototype, some rules of the design pattern from Mobile Design Pattern Gallery [28] are applied i.e.: menus, the use primary navigation and secondary navigation patterns. An example, when we design the form menu, we use the subject selection which rely on menu list mode. In the main menu form, the menu-list presents options of materials, quizes, and other features. Besides, it also uses icons in accordance with the name of the title which function aesthetically and enable users to understand and remember the menu presented (see Fig. 3). For the second primary navigation pattern, control tab is used to navigate and to present another page without accessing or opening another visual control. Control tab can ease users in accessing the main menu since it shows the options in the main menu of the application even though the users have moved to other pages.

In designing the content, secondary navigation is employed to expand list pattern. Expanding list is useful for managing and presenting information based on the information group that the users need to see. As a menu expanding list can be opened and closed so we can input the necessary information. To make easier, a sign of status, an arrow, is implemented. When the arrow faces down, it shows that the expanding list is closed, while facing up means it is open and showing the contained information. An arrow icon is also used to make easier. It helps users to go back to the previous pages. The title bar is given a label that informs the page being accessed by the users.

The evaluation form was designed in accordance with multiple choice evaluation pattern in general. There is only one question within one page and it is used to fit the

Fig. 3. Interface Design of prototype mobile learning in blue colour (The other element design i.e. colour and shape is also applied, this illustration is only one example of prototype using the blue colour and round shape) (Color figure online)

screen limitation of smartphone dimension. Radio control button is used to answer the question. Radio control button serves the function to choose one among some options provided. A "next" button is provided to present the following question, moreover it will enables users to choose their answer for the question. An arrow is provided on the left or upper left of the title bar which is used to move to the previous pages. The title bar also shows the location of the page being accessed by the users.

The result of experiment one i.e. the colours (blue, red and green) and round shape are potentials in generating positive emotion. So, this result will be implemented in the interface mobile learning prototype especially in the designs of background, foreground, and buttons. Monochromatic technique is used to combine the colours by mixing one colour with white or black. The more white is used the colour will be lighter while more black will bring darker colour. Ilustration 3 shows the detail of the mock-up of the model which use blue colour and round shape as the element design.

5.1 Data Collecting

Data were collected from 41 respondents who had been involved in the first experiment. The second experiment employed an android smartphone as a medium to run mobile learning application prototype and used SAM Questionnaire to find out the respondents' responses towards the stimuli. Data-collecting process is started when respondents operated m-learning application prototype. While the application was running, the respondents were asked to do five sequential activities. The first is to access three main menu in Tab Menu: Materials, Quizes, and Others. The second is to choose the material of menu list "PengenalanTeknologi Informasi". The third is to open the sub-material "Pengertian TI" and read the content presented. The fourth activity asks respondents to access quiz menu tab "Quiz Pengenalan Teknologi Informasi". The last activity required respondents to answer the multiple choice questions provided. Furthermore The data collection follow the Fig. 2 where the model is the prototype of m-learning.

5.2   Data Analysis and Results of Experiment Two

This stage employed OWA to analyze the data and to test the average mean of each mobile learning prototype model. Furthermore, OWA is used to find which prototype generates the positive, negative or neutral emotions. This experiment went through reliability test by using Alpha Cronbach technique towards three questions of prototype given to the 41 respondents. The reliability test which showed the score of Cronbach Alpha (0.812), was then compared with the r-table score for $df = 39$ which is 0.3081 (significance level 0.05), so it can be concluded that the questionnaire employed here is reliable since the score of Cronbach Alpha is (0.318) > r-$Table$ (0.3081). The OWA of thir experiment is listed in Table 7.

Tabel 7. Frequency analysis of prototype M-Learning

Prototype	Very negative		Negative		Neutral		Positive		Very positive		Total
	F	%	F	%	F	%	F	%	F	%	N (%)
Blue	0	0	0	0	7	17.1	17	41.5	17	41.5	41 (100)
Red	0	0	0	0	8	19.5	23	56.1	10	24.4	41 (100)
Green	0	0	8	19.5	19	46.3	13	31.7	1	2.4	41 (100)

Using Homogeneity of Variances we found that value $sig = 0{,}173$, since $sig > 0{,}05$ which means there are varians among prototye. Therefore the analysis is followed by OWA, and the result of OWA is showed in Table 8.

Table 8.   One-way anova analysis of prototype

	Sum of sq.	Df	Mean sq.	F	Sig.
Between groups	26.797	2	13.398	25.412	.000
Within groups	63.268	120	.527		
Total	90.065	122			

Using OWA analysis it is known that the value of $F_{measure}$ (25.412) is the compared with the value of $F_{table(2;120;0,05)}$ (3.07) it is found that $F_{measure} > F_{table}$ and the value of the comparison $sig(0.000) < 0.05$. It states that the three prototypes reach different emotional rating averages. The further observation, the average differences of the emotion rating scores of the three prototypes, is performed by LSD Post Hoc Test and is presented in Table 9.

As presented in Table 9, LSD post hoc test shows that the significant gaps in the rating average scores in mobile learning prototype happen between blue and green prototype, red and green prototypes. sig score for blue and green prototypesis sig (0.000) < 0.05 so it can be concluded that blue and green prototypes has different emotion rating mean scores. The same applies to red and green prototypes with sig score (0.000) < 0.05, so red and green prototypes has different emotion rating scores.

Table 9. Post hoc test LSD of three prototypes (with 95% Conf Int)

Colour 1	Colour 2	Mean Diff.	Std Err.	Sig	Low Bnd.	Upp Bnd.
Blue	Red	.19512	.16037	.226	−.1224	.5126
	Green	1.07317*	.16037	.000	.7556	1.3907
Red	Blue	−.19512	.16037	.226	−.5126	.1224
	Green	.87805*	.16037	.000	.5605	1.1956
Green	Blue	−1.07317*	.16037	.000	−1.3907	−.7556
	Red	−.87805*	.16037	.000	−1.1956	−.5605
Blue	Red	.19512	.16037	.226	−.1224	.5126
	Green	1.07317*	.16037	.000	.7556	1.3907

While blue and read prototypes has no significant difference in the emotion rating scores since the *sig* score is (0.226) > 0.05.

6 Discussion

From the colour model analysis, it is concluded that red, green and blue generate positive emotional state in respondents as there is no significant average gaps among the three models. However, when considering the average scores achieved by each colour, blue reaches 4.07372 (Positive) with the deviation standard of 0.87722 is the most effective in generating positive emotion, followed by red averaged in 4.0244 and deviation standard of 0.72415. While yellow tends to generate negative emotion.

The shape models conclude that there is no significant difference among the mean of emotion rating of each of the models. When considering the mean of the shape model, it is concluded that there are no differences among the emotion rating average in each shape model and when looking at the mean score, the round model with the average of 3.6098 potentially generates positive influence to the respondents. The three shape models tend to give neutral influence to the respondents emotion as shown in the average score that range between the scores of three to four.

Another conclusion from this experiment is that the blue colour and round shape tend to make positive emotion. The conclusion is represented by the average value has average value 3.8780 and standart deviation 0.92723. It is very contrast if we compare with green colour. From the analysis on the prototype mobile learning, it can be concluded that blue and red prototypes potentially generate positive emotions, proven by the inexistence of significant mean gap between the two prototypes. However, seen from the mean scores, blue is 4.2439 (Positive) and deviation standard of 0.73418 is the most effective colour that generates positive emotion followed by red with mean score of 4.0488 and deviation standard of 0.66900. On the other hand, green prototype tends to generate neutral emotion to the respondents.

7 Conclusion and Suggestion

The research on colour design element of red, green, blue and yellow, and shape design element of round, angular and combination of round and angular in experiment one shows results as follow. The colours in design element that tend to generate positive emotion are blue The shape design elements that generates positive emotion is round Experiment two proves that the combination of colours applied in visual of interface system generating positive emotion is blue and red prototypes.

To design an interface mobile learning it is necessary to consider emotional aspects based on the results of design elements of colours and shades that potentially generate positive emotions in experiment one. It is also necessary to apply a guideline of designing interface mobile application. The application of emotional aspects such as colours and shapes in designing interface mobile learning are in the interface components such as the background, foreground and buttons.

Acknowledgements. I would like to thank Atma Jaya University Yogyakarta, Indonesia and DIKTI for the financial support for my research project.

References

1. Charland, L.C.: Emotion as a natural kind: towards a computational foundation for emotion theory. Philos. Psychol. **8**(1), 59 (1995)
2. Whiting, D.: The feeling theory of emotion and the object-directed emotions. Eur. J. Philos. **19**(2), 281–303 (2011)
3. Fredrickson, B.L.: The role of positive emotions in positive psychology: the broaden-and-built theory of positive emotions. Am. Psychol. **56**(3), 218–226 (2001)
4. Park, B., Knörzer, L., Plass, J.L., Brünken, R.: Emotional design and positive emotions in multimedia learning: an eyetracking study on the use of anthropomorphisms. Comput. Educ. **86**, 30–42 (2015)
5. Kim, J., Lee, J., Choi, D.: Designing emotionally evocative homepages: an empirical study of the quantitative relations between design factors and emotional dimensions. Int. J. Hum Comput Stud. **59**(6), 899–940 (2003)
6. Nijdam, N.: Mapping emotion to colour. Emotion, pp. 2–9 (2007)
7. Naz, K., Epps, H.: Relationship between colour and emotion: a study of college students. Coll. Student J. 38, 396–406 (2004). (1968)
8. Adams, F.M., Osgood, C.E.: A cross-cultural study of the affective meanings of color. J. Cross Cult. Psychol. **4**(2), 135–156 (1973)
9. Hemphill, M.: A note on adults' color-emotion associations. J. Genet. Psychol. **157**(3), 275–280 (1996)
10. Sotgiu, I., Galati, D., Manzano, M., Gandione, M., Gómez, K., Romero, Y., Rigardetto, R.: Parental attitudes, attachment styles, social networks, and psychological processes in autism spectrum disorders: a cross-cultural perspective. J. Genet. Psychol. Res. Theor. Hum. Dev. **172**(4), 353–375 (2011). doi:10.1080/00221325.2010.544342
11. Gaines, K.S., Curry, Z.D.: The inclusive classroom: the effects of color on learning and behavior. J. Fam. Consum. Sci. Educ. **29**(1), 46–57 (2011)

12. Lu, X., Suryanarayan, P., Adams, R.B., Li, J., Newman, M.G., Wang, J.Z.: On shape and the computability of emotions. In: Proceedings of the 20th ACM İnternational Conference on Multimedia, pp. 229–238 (2012)
13. Holley, D., Sentence, S.: Mobile 'Comfort' zones: overcoming barriers to enable facilitated learning in the workplace. J. Interact. Media Educ. 2015(1)15 (2015). http://dx.doi.org/10.5334/jime.av
14. Sharples, M.: Mobile learning: research, practice and challenges. Distance Educ. China **3**(5), 5–11 (2013)
15. Abegaz, T.: Design with emotion: improving web search experience for older adults. all dissertations. Paper 1439 (2014)
16. Plass, J.L., Heidig, S., Hayward, E.O., Homer, B.D., Um, E.: Emotional design in multimedia learning: Effects of shape and color on affect and learning. Learn. Instr. **29**, 128–140 (2014)
17. Kumi, R., Conway, C., Limayem, M., Goyal, S.: Learning in color: how color and affect influence learning outcomes. IEEE Trans. Prof. Commun. **56**(1), 2–15 (2013)
18. Zimmermann, P., Guttormsen, S., Danuser, B., Gomez, P.: Affective computing–a rationale for measuring mood with mouse and keyboard. Int. J. Occup. Saf. Ergon. JOSE **9**(4), 539–551 (2003)
19. Gable, P., Harmon-Jones, E.: Approach-motivated positive affect reduces broadening of attention. Psychol. Sci. **19**, 476–482 (2008). http://pss.sagepub.com/content/19/5/476.short
20. Mahlke, S., Minge, M., Thüring, M.: Measuring multiple components of emotions in interactive contexts. In: CHI 2006 Extended Abstracts on Human Factors in Computing Systems - CHI 2006, pp. 1061–1066 (2006)
21. Joblove, G.H., Greenberg, D.: Color spaces for computer graphics. ACM SIGGRAPH Comput. Graph. **12**(3), 20–25 (1978)
22. Zhao, S., Gao, Y., Jiang, X., Yao, H., Chua, T., Sun, X.: Exploring principles-of-art features for image emotion recognition. In: Proceedings of the ACM International Conference on Multimedia, MM 2014, pp. 47–56 (2014)
23. Bradley, M., Lang, P.J.: Measuring emotion: the self-assessment semantic differential manikin and the. J. Behav. Ther. Exp. Psychiatry **25**(I), pp. 49–59 (1994)
24. Jacobs, K.W., Suess, J.F.: Effects of four psychological primary colors on anxiety state. Perceptual and motor skills (1975)
25. Valdez, P., Mehrabian, A.: Effects of color on emotions. J. Exp. Psychol. Gen. **123**(4), 394–409 (1994)
26. Watson, D., Clark, L.A, Tellegen, A.: Development and validation of brief measures of positive and negative affect: the PANAS scales. J. Pers. Soc. Psychol. **54**(6), 1063–70 (1988)
27. Dimberg, U., Thunberg, M., Elmehed, K.: Unconscious facial reactions to emotional facial expressions. Psychol. Sci. J. Am. Psychol. Soc. APS **11**(1), 86–89 (2000)
28. Neil, T.: Mobile Design Pattern Gallery. pp. 1–37. O'Reilly Media, Inc. (2012)

Author Index

Printed in the United States
By Bookmasters